Residential Construction Academy

Basic Principles for Construction

Residential Construction Academy

Basic Principles for Construction

Second Edition

Mark W. Huth

THOMSON

DELMAR LEARNING

Australia Canada Mexico Singapore Spain United Kingdom United States

THOMSON

™

DELMAR LEARNING

Residential Construction Academy: Basic Principles for Construction, Second Edition
Mark W. Huth

Vice President, Technology and Trades ABU:
David Garza

Director of Learning Solutions:
Sandy Clark

Managing Editor:
Larry Main

Senior Acquisitions Editor:
James DeVoe

Senior Product Manager:
John Fisher

Marketing Director:
Deborah S. Yarnell

Marketing Manager:
Kevin Rivenberg

Marketing Specialist:
Mark Pierro

Director of Production:
Patty Stephan

Production Manager:
Stacy Masucci

Content Project Manager:
Jennifer Hanley

Technology Project Manager:
Kevin Smith

Editorial Assistant:
Tom Best

Library of Congress Cataloging-in-Publication Data
Huth, Mark W.
  Residential construction academy : basic principles for construction / Mark W. Huth. — 2nd ed.
    p. cm.
  Includes bibliographical references and index.
    ISBN 1-4180-5251-5 (alk. paper)
1. Building—Textbooks. I. Title. II. Title: Basic principles for construction.
  TH146.H88 2007
  690'.8—dc22   2007011041

NOTICE TO THE READER

Publisher does not warrant or guarantee any of the products described herein or perform any independent analysis in connection with any of the product information contained herein. Publisher does not assume, and expressly disclaims, any obligation to obtain and include information other than that provided to it by the manufacturer.

The reader is expressly warned to consider and adopt all safety precautions that might be indicated by the activities herein and to avoid all potential hazards. By following the instructions contained herein, the reader willingly assumes all risks in connection with such instructions.

The Publisher makes no representation or warranties of any kind, including but not limited to, the warranties of fitness for particular purpose or merchantability, nor are any such representations implied with respect to the material set forth herein, and the publisher takes no responsibility with respect to such material. The publisher shall not be liable for any special, consequential, or exemplary damages resulting, in whole or part, from the readers' use of, or reliance upon, this material.

Table of Contents

Preface

Home Builders Institute Residential Construction Academy: Basic Principles for Construction

About the Residential Construction Academy Series

One of the most pressing problems confronting the building industry today is the shortage of skilled labor. It is estimated that the construction industry must recruit 200,000 to 250,000 new craft workers each year to meet future needs. This shortage is expected to continue well into the next decade because of projected job growth and a decline in the number of available workers. At the same time, the training of available labor is becoming an increasing concern throughout the country. This lack of training opportunities has resulted in a shortage of 65,000 to 80,000 skilled workers per year. The crisis is affecting all construction trades and is threatening the ability of builders to build quality homes.

These are the reasons for the creation of the innovative *Residential Construction Academy Series*. The *Residential Construction Academy Series* is the perfect way to introduce people of all ages to the building trades while guiding them in the development of essential workplace skills including carpentry, electrical, HVAC, plumbing, and facilities maintenance. The products and services offered through the *Residential Construction Academy* are the result of cooperative planning and rigorous joint efforts between industry and education. The program was originally conceived by the National Association of Home Builders (NAHB)—the premier association of over 200,000 member groups in the residential construction industry—and its workforce development arm, the Home Builders Institute (HBI).

Construction professionals and educators created National Standards for the construction trades. In the summer of 2001, the NAHB, through the HBI, began the process of developing residential craft standards in five trades. They are carpentry, electrical wiring, HVAC, plumbing, and facilities maintenance. Groups of construction employers from across the country met with an independent research and measurement organization to begin the development of new craft-training standards. The guidelines from the National Skills Standard Board were followed in developing the new standards. In addition, the process met or exceeded the American Psychological Association standards for occupational credentialing.

Then, through a partnership between HBI and Delmar Learning, learning materials—textbooks, videos, and instructor's curriculum and teaching tools—were created to effectively teach these standards. A foundational tenet of this series is that students *learn by doing*. A constant focus of the *Residential Construction Academy* is teaching the skills needed to be successful in the construction industry and constantly applying the learning to real-world applications.

Perhaps most exciting to learners and industry is the creation of a National Registry of students who have successfully completed courses in the *Residential Construction Academy Series*. This registry or transcript service provides an opportunity for easy access for verification of skills and competencies achieved. The Registry links construction industry employers and qualified potential employees together in an online database, facilitating student job search and the employment of skilled workers. For more information on the *Residential Construction Academy Series,* visit http://www.residentialacademy.com

About This Book

Whether an individual chooses a career as a skilled craftsperson or is striving to become a general contractor, *Basic Principles for Construction* provides the necessary background for understanding the construction industry and the basic skills for learning a specific trade.

Basic Principles for Construction is an outstanding resource for new and advancing construction students, or for those considering entering a construction program. This text provides a solid foundation to learn the major trade areas—carpentry, electrical wiring, HVAC, plumbing, and facilities maintenance. It introduces students to the industry—explaining how it is organized and how to successfully gain employment—and also covers the need-to-know information for the daily activities associated with working in the industry, including safety, basic math, tools, and blueprint reading.

Organization

This textbook is organized in a logical sequence that is easy to learn and teach, and is divided into five major sections:

- **Section 1: The Construction Industry** is designed to provide students with background on the industry. It introduces students to the organization and leadership structure, as well as the importance of ethics, teamwork, and effective communications with others to successfully complete a job.
- **Section 2: Safety** covers the all-important elements of safely working on a job site, including working with electricity, hazardous materials, scaffolding, ladders, and compressed air. It also covers OSHA regulations, practical housekeeping, and personal protective equipment to ensure safe work habits.
- **Section 3: Construction Math** reviews basic math skills and how to practically apply these skills on the job. Examples and practice problems are integrated into the chapters to increase student aptitude in working out various construction problems. The section concludes with a chapter on combined operations, which illustrates the necessity of having the ability to utilize several math skills in completing a single job.
- **Section 4: Tools and Fasteners** introduces students to the selection, use, and care of the various hand and power tools required to complete a job, as well as different types of fasteners. The section also includes a chapter on the basics of rigging as it applies to residential construction.
- **Section 5: Print Reading** emphasizes the elements and features of basic residential blueprints and how to accurately read them. Activities at the end of each chapter encourage students to practice their blueprint reading skills.

New to This Edition

The second edition continues the structure, organization, and all of the content that made the first edition so successful; but in this edition we added content that some instructors thought was missing from the first edition. Specifically, we added coverage of building codes, trade unions, and nonunion organizations; material handling was added to the safety coverage; and a new chapter was added on percent (percentage).

We encourage instructors and students to comment on what they like about this book and what they would like to see improved. Comments from those who use the text are our most valuable resource for continuing to make *Basic Principles for Construction* the best it can be be.

Features

This innovative series was designed with input from educators and industry and informed by the curriculum and training objectives established by the Standards Committee. The following features aid learning:

Success Stories

Donna Story

TITLE
President and founder, U.S. Plumbing in Charlotte, Tennessee

EDUCATION
Donna completed high school and holds an associate's degree in computer accounting. She has also completed the plumber's state licensing exam in Tennessee.

HISTORY
Donna quickly realized that a desk job in computer accounting was not for her. She took her first plumbing job with H & H Plumbing in Nashville. In 2001, she started a new business with her husband, Steve.

ON THE JOB
Donna and Steve work about 10 hours a day, starting with an hour in the office each morning to set up the day's work. They divide the site work between them. Donna reports: "I run copper for water lines and do the soldering. Steve runs plastic for drain lines." They complete billing and other paperwork at night. Donna hires and manages a few high school students during the summer.

BEST ASPECTS
"It's always rewarding to see the finished product," Donna reflects. "I would never have dreamed that so much work goes into plumbing a house. When homeowners are proud and pleased, that's a big reward."

CHALLENGES
Donna describes the physical challenge of working in crawl spaces: "Residential work is a little more challenging than commercial because you're often lying on your back all day under a house" in cramped conditions. She has also worked hard to build confidence in a male-dominated field. "A lot of women out there don't like office jobs," Donna notes. "They're outside people. But they're so intimidated by the construction field. I'm not easily intimidated."

IMPORTANCE OF EDUCATION
"It takes a lot of math skill to read blueprints," states Donna, who values her success in high school math courses. College courses in psychology have helped her learn to work with people. "You're often working directly with the customer," she says, reminding that "on a remodel, you're working in their living quarters. You must have good people skills."

FUTURE OPPORTUNITIES
Within 5 to 10 years, Donna wants to employ and supervise about five people. She hopes to plumb grocery stores one day, and will consider earning a general contractor's license.

WORDS OF ADVICE
"Plumbing is a wonderful field with a good living to be made. See if you like it. If you do, then go for the gusto. The sky's the limit."

A **Success Story** opens each section, providing insights, advice, and motivation from professionals working in a variety of construction trades, and offering an insider's view of construction as a career.

Chapter 3 Jobsite Safety

OBJECTIVES

After completing this chapter, the student should be able to:

- explain what an accident is and what causes accidents.
- define OSHA and explain its impact on construction workers.
- identify the appropriate PPE for common work situations and explain how to use it.
- explain the Hazard Communication Standard and find information on a Material Safety Data Sheet.
- explain how fires are ignited, sustained, and extinguished.
- recognize the dangers of working in trenches and explain how to work safely in a trench.
- explain electric shock and list safety considerations in working around electricity.

Learning Features such as the **Objectives** and **Glossary of Terms** set the stage for the coming body of knowledge and help the learner identify key concepts and information. These learning features serve as a road map throughout each chapter and offer a practical resource for reference and review.

The **Glossary** at the end of the book is now written in both English and Spanish.

Glossary

amount principal plus interest.

cantidad principal más interés.

ampere the unit of measure for electric current. Also abbreviated amp. Many power tool motors are sized according to the amperage their motor draws.

amperio la unidad de la medida para la corriente eléctrica. También abreviado amp. Muchos motores de la herramienta eléctrica se clasifican según el amperaje que su motor dibuja.

anchor a device that can be driven or set in concrete, masonry, or other material to provide a place to attach a bolt. There are several types of anchors.

ancla un dispositivo que se pueda conducir o fijar en el concreto, la albañilería, o el otro material para proporcionar un lugar a la fijación un perno. Hay varios tipos de anclas.

apprentice a person who is being trained to work in the building trades. Apprentices attend classes and work under the supervision of a skilled craftsman.

aprendiz una persona que se está entrenando para trabajar en los comercios de edificio. Los aprendices atienden a clases y al trabajo bajo supervisión de un artesano experto.

area the space inside a shape.

area el espacio dentro de una forma.

ball bearing a style of bearing in which moving parts roll on steel balls.

rodamiento de bolitas un estilo de llevar en que piezas móviles ruedan en las bolas de acero.

base of a triangle the side opposite the corner from which the height is measured. This can be any side of the triangle.

base de un triángulo el lateral enfrente de la esquina de la cual se mide la altura. Éste puede ser cualquier lado del triángulo.

blade (rafter square) the longer arm of a square.

lámina (cuadrado del viga) el brazo más largo de un cuadrado.

body language unwritten, unspoken message communicated by the way we hold our body or the expressions on our face.

la lengua de cuerpo el mensaje no escrita y no hablado que nos comunica de la manera llevamos nuestro cuerpo o de la expresión de la cara.

box nail has a thin shank like a finishing nail, but a flat head like a common nail. Usually coated to prevent loosening.

el clavo de la caja tiene una caña fina como un clavo que acaba, pero una cabeza plana como un clavo común. Generalmente cubierto para evitar aflojamiento.

brad a very short nail with a small head, used to fasten thin parts.

alfilercillo un clavo muy corto con una cabeza pequeña, usada para sujetar piezas finas.

cap screw a small bolt, usually with a hexagonal head.

tornillo de casquillo un perno pequeño, generalmente con una cabeza hexagonal.

carriage bolt a large bolt for use in wood. Has a smooth oval head and a section of square shank right below the head.

perno de carro un perno grande para el uso en madera. Tiene una cabeza oval lisa y una sección de la derecha

base de un triángulo el lateral cuadrada de la caña debajo de la cabeza.

centerline a line used to indicate the center axis of an object.

línea central una línea para indicar el eje de centro de un objeto.

chuck the part of a drill that holds the drill bit.

arroje la pieza de un taladro que celebre el pedacito de taladro.

chuck key a special tool used to tighten a drill chuck.

llave de Arroje una herramienta especial usada para apretar una tirada del taladro.

circle a shape in which every point on the perimeter is the same distance from a center point.

círculo una forma en la cual cada punto en el perímetro es a la misma distancia de un punto de centro.

class A fire a fire that involves ordinary materials like paper, cardboard, and wood. Class A fires can be extinguished with water.

fuego de la clase A un fuego que implica los materiales ordinarios como el papel, la cartulina, y la madera. Los fuegos de la clase A se pueden extinguir con agua.

class B fire a fire that involves flammable liquids. Class B fires are extinguished with either dry chemicals or CO$_2$ (carbon dioxide).

el fuego de la clase B un fuego que implique líquidos inflamables. Los fuegos de la clase B se extinguen con los productos químicos secos o el CO$_2$ (bióxido de carbono).

class C fire an electrical fire. Class C fires are extinguished with CO$_2$ (carbon dioxide).

Caution features highlight safety issues and present urgent safety reminders for those working with the various tools in industry—so that students can avoid potential mishaps.

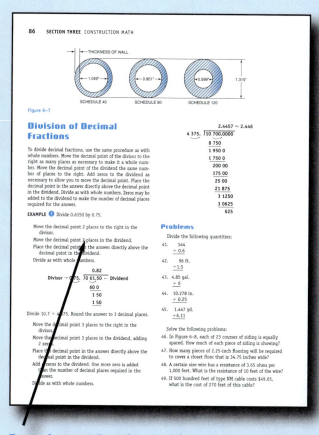

Examples and **Practice Problems** integrated into the math chapters illustrate for students, step by step, the various methods of working out construction problems. In addition, it encourages them to practice and improve their math skills.

Review Questions are designed to reinforce the information in the chapter as well as give students the opportunity to think about what has been learned and what they have accomplished.

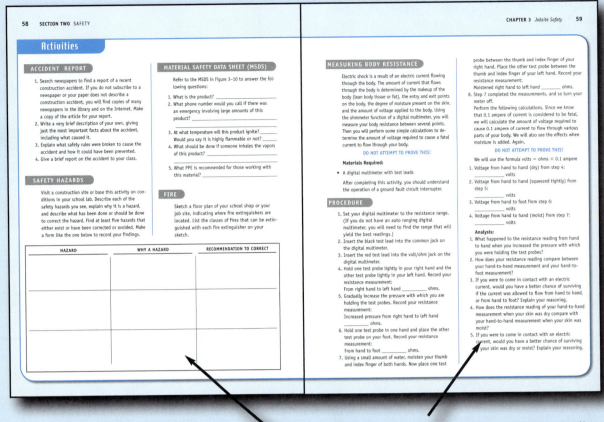

Activities complete each chapter where applicable, and are intended to provide students with a practical "hands-on" experience as it relates to the reading within the text. Everything from identification of tools to Internet research, critical thinking, and building—these activities bring the key points of the chapter to life!

Turnkey Curriculum and Teaching Material Package

We understand that a text is only one part of a complete, turnkey educational system. We also understand that instructors want to spend their time on teaching, not preparing to teach. The *Residential Construction Academy Series* is committed to providing thorough curriculum and preparatory materials to aid instructors and alleviate some of those heavy preparation commitments. An integrated teaching solution is ensured with this text, the *Instructor's e.resource*™, and the print *Instructor's Resource Guide*.

e.resource™

Delmar Learning's **e.resource**™ is a complete guide to classroom management. Designed as an integrated package, the e.resource offers the Instructor many valuable tools, including **PowerPoint**, a **Computerized Testbank,** and an **Image Library,** as well as the chapter outlines and answers to review questions available through the electronic version of the print **Instructor's Resource Guide.**

PowerPoint

The series includes a complete set of PowerPoint presentations providing lecture outlines that can be used to teach the course. Instructors may teach from this outline or can make changes to suit individual classroom needs.

Computerized Testbank

The Computerized Testbank contains hundreds of questions that can be used for in-class assignments, homework, quizzes, or tests. Instructors can edit the questions in the testbank, or create and save new questions.

Image Library

An Image Library offers instructors the option of creating their own classroom presentations by providing electronic versions of all line art and photos from the textbook.

Instructor's Resource Guide

An instructor's version of lecture outlines, the Instructor's Resource Guide provides a step-by-step breakdown of the key points found in each chapter, along with "Teaching Tips" and correlating PowerPoint presentation slides, creating a completely streamlined and integrated approach to teaching. Also included are answers to the Review Questions that appear at the end of each chapter. The Instructor's Resource Guide is available in electronic and print versions.

Workbook

Designed to accompany *Residential Construction Academy: Basic Principles of Construction, Second Edition,* this Workbook is an extension of the core text, and provides additional review questions and problems designed to challenge and reinforce the student's comprehension of the content presented in the core text.

DVD

The *Residential Construction Academy: Basic Principles for Construction DVD* consists of four 20-minute videos that contain essential information covered in the text. It begins with safety, then covers math and measurement, hand and power tools, views, and elevations.

Online Companion

The Online Companion is an excellent supplement for students that features many useful resources to support the *Basic Principles for Construction* book. Linked from the Student Materials section of http://www.residentialacademy.com, the Online Companion includes chapter quizzes, an online glossary, additional "Success Stories," related links, and more.

The Complete Residential Construction Academy Series

Basic Principles of Construction provides a foundation for other texts within the series, which covers carpentry, electrical wiring, HVAC, plumbing, masonry, and facilities maintenance. Each title offers a complete instructor curriculum package, including accompanying videos and a CD-ROM courseware series. Programs in these trades may be credentialed by the Home Builders Institute, providing national recognition for the program. In addition, students who successfully complete one or more of the trade programs can receive a certification of completion and may be eligible to enter in the National Registry. This registry provides a direct link between students and potential employers. For applications and the latest information, visit http://www.residentialacademy.com

Available

Residential Construction Academy Carpentry (Order # 1-4018-1343-7)

Residential Construction Academy Electrical Principles (Order # 1-4018-1294-5)

Residential Construction Academy House Wiring (Order # 1-4018-1371-2)

Residential Construction Academy HVAC (Order # 1-4018-4899-0)

Residential Construction Academy Plumbing (Order # 1-4018-4891-5)

Coming Soon

Summer 2007: Residential Construction Academy: Masonry, Brick and Block Construction (Order # 1-4180-5284-1)

Summer 2007: Residential Construction Academy: Carpentry, 2E (Order # 1-4180-0183-X)

Fall 2007: Residential Construction Academy: Facilities Maintenance (Order # 1-4018-6483-X)

Fall 2007: Residential Construction Academy: House Wiring, 2E (Order # 1-4180-1098-7)

About the Author

The author of this textbook, Mark Huth, brings many years of experience in the industry to his writing—first working as a carpenter, contractor, a building construction teacher, and then as a publisher of construction textbooks. His career has allowed him to consult with hundreds of construction educators in high schools, colleges, and universities. *Basic Principles for Construction* has been shaped by his observations of the difficulties students have in studying construction, and by the outstanding programs offered at the best schools in the country. He has also authored several other successful construction titles, including *Construction Technology*, *Basic Blueprint Reading for Construction*, *Understanding Construction Drawings*, and *Practical Problems in Mathematics for Carpenters*.

Acknowledgments

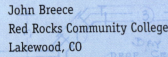any experts within the field contributed their time and expertise to the project. The National Association of Home Builders, Home Builders Institute, Delmar Learning, and the author extend our sincere appreciation to:

John Breece
Red Rocks Community College
Lakewood, CO

Shannon Brown
Minico High School
Rupert, ID

Paul Drake
Red Rock Job Corps Center
Lopez, PA

Joseph Dusek
Triton College
River Grove, IL

Earl Garrick
Pima Community College
Tucson, AZ

David Gehlauf
Tri-County Vocational School
Nelsonville, OH

Larry Kness
Southeast Community College
Milford, NE

Daryl Martinez
Job Corps
Tahlequah, OK

Mark Martin
HBI Penobscot Job Corp Center
Bangor, MA

Steve Miller
Construction Technology Consultant
North Carolina Department of Education
Raleigh, NC

Ed Moore
York Technical College
Rock Hill, SC

Lee Morris
Construction Education Services
Griffin, GA

Bruce Purdy
Grove City High School
Grove City, OH

Deanne Robertson
HBI Project Coordinator
Colorado Springs, CO

David Robinson
LA Trade Technical College
Los Angeles, CA

Merl Rogers
LA Trade Technical College
Los Angeles, CA

Les Stackpole
Eastern Maine Technical College
Bangor, ME

Kevin Ward
McEachern High School
Powder Springs, GA

The publisher and author also wish to express a special thanks to Mike Brumbach of York Technical College in Rock Hill, SC, and Barry Burkan of Apex Technical College in New York, NY, for their ideas and contributions to the writing of the text. A special thanks also to Mary Clyne for her contribution to the "Success Stories" that appear in this text and in the accompanying Online Companion.

The Construction Industry

Success Stories

Sal Inguanti

The amount of paperwork increases with every job. One of the biggest challenges on the job is time—time is money! Sal is working hard to stay organized and to maintain schedules. "Scheduling is important—as is making sure my subcontractors stay on schedule, too."

TITLE

Owner/partner, P & I Custom Builders, Westchester County, New York

EDUCATION

Sal graduated from a two-year carpentry program at Southern Westchester Board of Cooperative Educational Services (always called BOCES—this is a vocational program offered to all public schools in the county). He holds an associate's degree in carpentry from State University of New York, Delhi College.

HISTORY

Sal lives and works in Westchester County, New York. Growing up, he always enjoyed working with his hands and spent a lot of time working with his grandfather, where he learned a little bit of everything. Sal worked for a contractor for two years after graduating from college, and then worked for a builder for almost two years building spec (speculative) homes. Sal was a lead carpenter for both of these companies. His company builds high-end spec homes and additions, and does remodeling, bathrooms, cabanas, flood restoration, and drainage work.

CHALLENGES

The transition from being a lead carpenter to owning his own business and being a foreman for the whole job was tough. "I find myself using tools less and less. I'm responsible for supervising all of the subcontractors on my job and for the work that they do."

IMPORTANCE OF EDUCATION

"Everything I learned at Delhi College has really set me apart, and people are amazed at what I know at my age. One of my instructors always said, 'Math is money!' He was right, but math applies to everything. The more you learn in the classroom, the easier your work will be in the field. Diversify yourself. I'm able to do 90 percent of the work my subcontractors do when necessary."

FUTURE OPPORTUNITIES

Right now the sky's the limit! His company has just completed its biggest project to date, an 11,000-square-foot house in Rye, New York, which was listed for $4.8 million. He and his partner ran the job—from knocking down the existing house to creating the final landscaping. Sal's company has completed three spec projects and is now working on an 8,000-square-foot multimillion dollar custom home for a client. Sal and his partner have purchased another parcel of land and are working with an architect to design a home, and they are negotiating two other projects as well. "I worked hard to get to the point where I started investing my money and sweat equity into the projects we're currently doing. It gives me the sense of satisfaction that I'm not just swinging a hammer; I'm creating my own little 'pot of gold' at the end of each job."

WORDS OF ADVICE

"Learn every possible thing you can about construction and you'll never go hungry."

Chapter 1 | Organization of the Industry

OBJECTIVES

After completing this chapter, the student should be able to:

- ❂ list and describe several potential careers in construction.
- ❂ explain the roles of architects, engineers, city building officials, and contractors.
- ❂ describe the major forms of business ownership and the differences between them.
- ❂ explain what a building code is.

Glossary of Terms

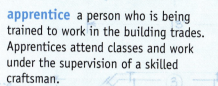

apprentice a person who is being trained to work in the building trades. Apprentices attend classes and work under the supervision of a skilled craftsman.

contractor the person who owns the construction business. Contractors enter into contracts with customers to do specified construction work. Contractors hire workers or other subcontractors to complete the contracted work.

corporation a form of business ownership in which people who are not involved in operating the business own shares of the company. The company is operated by a board of directors.

craft see **skilled trades.**

craft union members of a particular craft who are organized to work for the betterment of all members of the group. Union members pay dues as a requirement of their membership.

developer the person or company that buys undeveloped land and works with architects and contractors to develop it into more valuable property.

journeyman a skilled craft worker who has completed an apprenticeship or otherwise proved his or her ability in the trade.

laborer an unskilled or semiskilled worker on a construction site.

model code a suggested building code, intended to be adopted as is or with revisions to become an official code of a particular government.

partnership a form of business in which more than one person shares the ownership and operating duties for a company.

profession an occupation that requires more than four years of college and a license to practice.

semiskilled labor workers with very limited training or skills in the construction trades.

skilled trades the building trades—carpenters, electricians, plumbers, painters, and so on. These occupations require training and skill. The skilled trades are often referred to as the crafts.

sole proprietorship a business whose owner and operator are the same person.

subcontractor a contractor who is performing work for another contractor.

technicians technicians provide a link between the skilled trades and the professions by using mathematics, computer skills, specialized equipment, and knowledge of construction.

unskilled labor workers with no specific training in the construction trades. This term also applies to work that does not require training.

The residential construction industry is one of the biggest sectors of the American economy. According to *Engineering News Record,* a major construction news publication, one out of every six people is involved in construction in some way. The Home Builder's Institute reports that home building accounts for 52 percent of the construction industry. There are opportunities for people to work at all levels in the construction industry, from those who handle the tools and materials on the job site to the senior engineers and architects who spend most of their time in offices. Few people spend their entire lives in a single occupation, and even fewer spend their lives working for one employer. You should be aware of all the opportunities in the construction industry so that you can make career decisions in the future, even if you are sure of what you want to do at this time.

Construction Personnel

The occupations in the construction industry can be divided into four categories:

- unskilled or semiskilled labor
- skilled trades or crafts
- technicians
- design and management

Unskilled or Semiskilled Labor

Construction is labor intensive. That means it requires a lot of labor to produce the same dollar value of end products by comparison with other industries, where labor may be a smaller part of the picture. Construction workers with limited skills are called **laborers.** Laborers are sometimes assigned the tasks of moving materials, running errands, and working under the close supervision of a skilled worker. Their work is strenuous, and so construction laborers must be in excellent physical condition.

Construction laborers are construction workers who have not reached a high level of skill in a particular trade and are not registered in an apprenticeship program. These laborers often specialize in working with a particular trade, such as mason's tenders or carpenter's helpers (Fig. 1–1). Although the mason's tender may not have the skill of a bricklayer, the mason's tender knows how to mix mortar for particular conditions, can erect scaffolding, and is familiar with the bricklayer's tools. Many laborers go on to acquire skills and become skilled workers. Laborers who specialize in a particular trade are often paid slightly more than completely unskilled laborers.

Skilled Trades

A **craft** or **skilled trade** is an occupation working with tools and materials and building structures. The building trades are the crafts that deal most directly with building construction (Fig. 1–2).

Figure 1–1 This construction laborer is a carpenter's helper.

Carpenter
 Framing carpenter
 Finish carpenter
 Cabinetmaker
Plumber
 New construction
 Maintenance and repair
Roofer
Electrician
 Construction electrician
 Maintenance electrician
Mason
 Bricklayer (also lays concrete blocks)
 Cement finisher
HVAC technician
Plasterer
 Finish plaster
 Stucco plaster
Tile setter
Equipment operator
Drywall installer
 Installer
 Taper
Painter

Figure 1–2 Building trades.

The skill needed to be employed in the building trades is often learned in an **apprentice** program. Apprenticeships are usually offered by trade unions, trade associations, technical colleges, and large employers. Apprentices attend class a few hours a week to learn the necessary theory. The rest of the week they work on a job site under the supervision of a **journeyman** (a skilled worker who has completed the apprenticeship and has experience on the job). The term "journeyman" has been used for decades and probably will continue to be used for many more decades, but it is worth noting that many highly skilled building trades workers are women. Apprentices receive a much lower salary than journeymen, often about 50 percent of what a journeyman receives. The apprentice wage usually increases as stages of the apprenticeship are successfully completed. By the time the apprenticeship is completed, the apprentice can be earning as much as 95 percent of what a journeyman earns. Many apprentices receive college credit for their training. Some journeymen receive their training through school or community college and on-the-job training. In one way or another, some classroom training and some on-the-job supervised experience are usually necessary to reach journeyman status. Not all apprentice programs are the same, but a typical apprenticeship lasts 4 or 5 years and requires 144 hours per year of classroom training and 2,000 hours per year of supervised work experience.

The building trades are among the highest paying of all skilled occupations. However, work in the building trades can involve working in cold conditions in winter or blistering sun in the summer. Also, job opportunities will be best in an area where a lot of construction is being done. This should not be much of a threat to a person interested in a career in the trades. The construction industry is growing at a high rate nationwide. Generally plenty of work is available to provide a comfortable living for a good worker.

Technicians

Technicians provide a link between the skilled trades and the professions. Technicians often work in offices, but their work also takes them to construction sites. Technicians use mathematics, computer skills, specialized equipment, and knowledge of construction to perform a variety of jobs. Figure 1–3 lists several technical occupations.

Most technicians have some type of college education, often combined with on-the-job experience, to prepare them for their technical jobs. Community colleges often have programs aimed at preparing people to work at the technician level in construction. Some community college programs are intended especially for preparing workers for the building trades, while others have more of a construction management focus. Construction management courses, such as those listed in Figure 1–4 give the graduate a good overview of the business of construction. The starting salary for a construction technician is about the same as for a skilled trade, but the technician can be more certain of regular work and will have better opportunities for advancement.

Design and Management

Architecture, engineering, and contracting are the design and management professions. The **professions** are those occupations that require more than four years of college and

Technical Career	Some Common Jobs
Surveyor	Measures land, draws maps, lays out building lines, and lays out roadways
Estimator	Calculates time and materials necessary for project
Drafter	Draws plans and construction details in conjunction with architects and engineers
Expeditor	Ensures that labor and materials are scheduled properly
Superintendent	Supervises all activities at one or more job sites
Inspector	Inspects project for compliance with local building codes at various stages of completion
Planner	Plans for best land and community development

Figure 1–3 Technicians.

a license to practice. Many contractors have less than four years of college, but they often operate at a very high level of business, influencing millions of dollars, and so they are included with the professions here. These construction professionals spend most of their time in offices and are not frequently seen on the job site.

Architects usually have a strong background in art, so they are well prepared to design attractive, functional buildings. A typical architect's education includes a four-year degree in fine art, followed by a master's degree in architecture. Most of their construction education comes during the final years of work on the architecture degree.

Engineers generally have more background in math and science, so they are prepared to analyze conditions and calculate structural characteristics. There are many specialties within engineering, but civil engineers are the ones most commonly found in construction. Some civil engineers work mostly in road layout and building. Other civil engineers work mostly with structures in buildings. They are sometimes referred to as structural engineers.

Contractors are the owners of the businesses that do most of the building. In larger construction firms, the principal (the owner) may be more concerned with running the business than with supervising construction. Some contractors are referred to as general contractors and others as **subcontractors** (Fig. 1–5). The general contractor is the principal construction company hired by the owner to construct the building. A general contractor might have only a skeleton crew, relying on subcontractors for most of the actual construction. The general contractor's superintendent coordinates the work of all the subcontractors.

First Year First Semester Course #	Title	Credit Hrs.
FORM 101	College Forum	1
CIVL 114	Construction Materials	2
CNST 100	Construction Surveying	3
CNST 170	Blueprint Reading	2
ENGL 101	English Composition I	3
MATH 150	College Algebra & Trigonometry	4
	Humanities or Social Science Elective	3
	Semester Total	**18**

Second Semester Course #	Title	Credit Hrs.
CNST 110	Statics and Strength of Materials	3
CNST 120	Architectural Drawing I	2
CNST 130	Principles and Practices of Light Construction	3
ENGL 106	English Composition II: Writing for Technicians	3
ACTG 100	Applied Accounting	3
MATH 151	Analytic Geometry & Basic Calculus	4
	Semester Total	**18**

Second Year First Semester Course #	Title	Credit Hrs.
CNST 230	Construction Management Sem.	3
CNST 270	Soils in Construction	3
CNST 220	Architectural Drawing II	3
CNST 210	Steel Construction	3
CNST 102	Construction Estimating	3
PHYS 115	Physics	4
	Semester Total	**19**

Second Semester Course #	Title	Credit Hrs.
CNST 231	Building Service Systems	3
CNST 211	Concrete Construction	3
CNST 232	Site Development	3
CNST 202	Construction Planning & Control	3
CNST 239	Construction Capstone	3
	Semester Total	**15**
	Total Credits Required	**70**

Figure 1–4 Construction management program at a community college.

It is quite common for a successful journeyman to start his or her own business as a contractor, specializing in the field in which he or she was a journeyman. These are the subcontractors that sign on to do a specific part of the construction, such as framing or plumbing. As the contractor's company grows and the company works on several projects at one time, the skilled workers with the best ability to lead others may become foremen. A foreman is a working supervisor of a small crew of workers in a specific trade. All contractors have to be concerned with business management. For this reason, many successful contractors attend college and get a degree in construction management. Most states require contractors to have a license to do contracting in their state. Requirements vary from state to state, but a contractor's license usually requires several years of experience in the trade and a test on both trade information and the contracting business.

An Overall View of Design and Construction

To understand the relationships between some of the design and construction occupations, we shall look at a typical housing development. The first people to be involved are the community planners and the real estate **developer.** The real estate developer has identified a 300-acre tract on which he would like to build nearly 1,000 homes, which he will later sell at a good profit. The developer must work with the city planners to ensure that the use he has planned is acceptable to the city. The city planner is responsible for ensuring that all building in the city fits the city's development plan and zoning ordinances. On a project this big, the developer might even bring in a planner of his own to help decide where parks and community buildings should be located and how much parking space they will need.

As the plans for development begin to take shape, it becomes necessary to plan streets and to start designing houses to be built throughout the development. A civil engineer is hired to plan and design the streets. The civil engineer will first work with the developer and planners to lay out the locations of the streets, their widths, and drainage provisions to get rid of storm water. (Did you ever consider how much water falls on a 1-mile-long by 32-foot-wide street when an inch of rain falls? More than 105,000 gallons! Where does that water go?) The civil engineer also considers soil conditions and expected traffic to design the foundation for the roadway.

An architectural firm, or perhaps a single architect, will design the houses. Typically several stock plans are used throughout a development, but many homeowners wish to pay extra to have a custom home designed and built. In a custom home, everything is designed for that particular house. Usually the homeowner, who will eventually live in the house, works with the architect to specify the sizes, shapes, and locations of rooms, interior and exterior trim, type of roof, built-in cabinets and appliances, use of outdoor spaces, and other special features. Architects specialize in use of space, aesthetics (attractive appearance), and livability features. Most architectural features do not involve special structural considerations, but when they do, a structural engineer is employed to analyze the structural requirements

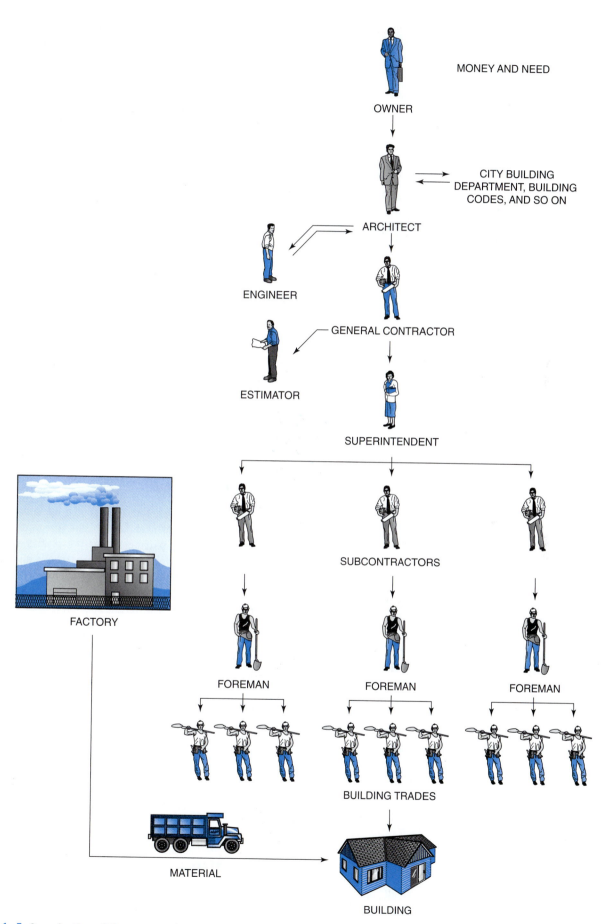

MONEY AND NEED

OWNER

CITY BUILDING
DEPARTMENT, BUILDING
CODES, AND SO ON

ARCHITECT

ENGINEER

GENERAL CONTRACTOR

ESTIMATOR

SUPERINTENDENT

SUBCONTRACTORS

FACTORY

FOREMAN

FOREMAN

FOREMAN

BUILDING TRADES

MATERIAL

BUILDING

Figure 1–5 Organization of the construction industry.

Figure 1–6 Trusses are designed by engineers.

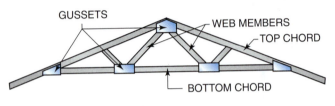

Figure 1–7 Parts of a roof truss.

and help ensure that the structure will adequately support the architectural features.

One part of construction that almost always involves an engineer is the design of roof trusses. Roof trusses are the assemblies that make up the frame of the roof (Fig. 1–6). Trusses are made up of the top chords, bottom chords, web members, and gussets (Fig. 1–7). The engineer considers the weight of the framing materials, the weight of the roof covering, the anticipated weight of any snow that will fall on the roof in winter, and the span (the distance between supports) of the truss to design trusses for a particular purpose. The architect usually hires the engineer for this work, and so the end product is one set of construction drawings that includes all the architectural and engineering specifications for the building. Even though the drawings are sometimes referred to as architectural drawings, they include work done by architects, engineers, and their technicians. Building codes require an architect's seal on the drawings before work can begin. The architect will require an engineer to certify certain aspects of the drawings before putting the architect's seal on them.

Forms of Ownership

Construction companies vary in size from small, one-person companies to very large international organizations that do many kinds of construction. However, the size of the company does not necessarily indicate the form of ownership.

Sole Proprietorship

The **sole proprietorship** is the easiest form of ownership to understand. The two words in the name of this form clearly describe it. Sole means only one or single. The proprietor of a business is the owner and operator. So a sole proprietorship is a business whose owner and operator are the same person. Sole proprietor construction companies are usually small companies in which the owner is one of the main workers.

Entrepreneurs are often sole proprietors. An entrepreneur is someone who starts a small business, often taking considerable financial risk. Small entrepreneurs started many of the largest, most successful businesses in the world today. The keys to successful entrepreneurship are understanding (not necessarily eliminating) the risks and doing thorough planning.

Each form of business ownership has advantages and disadvantages. The advantages of the sole proprietorship are that the owner has complete control over the business and that there is a minimum of government regulation. If the company is successful, the owner receives high profits. However, if the business goes into debt, the owner is responsible for that debt. The owner can be sued for the company, and the owner suffers all the losses of the company.

Partnership

A **partnership** is similar to a sole proprietorship, but there are two or more owners, rather than just one. In a general partnership, each partner shares the profits and losses of the company in proportion to the partner's share of investment in the company. General partnerships are common among engineering and architectural companies where each partner is an expert in a different specialty.

In a general partnership, each partner can be held responsible for all the debts of the company. The advantage of this form of ownership is that the partners share the expense of starting the business. Also partnerships, like sole proprietorships, are not controlled by extensive government regulations.

A variation of the general partnership is the limited liability partnership (LLP). A limited liability partner is one who invests in the business, receives a proportional share of the profit or loss, but has limited liability. In other words, a limited liability partner can only lose his or her investment. Every LLP must have one or more general partners who run the business. The general partners in an LLP have unlimited liability. They can be personally sued for any debts of the company.

Corporation

In a **corporation** a group of people own the company. Another, usually smaller, group of people manage the business.

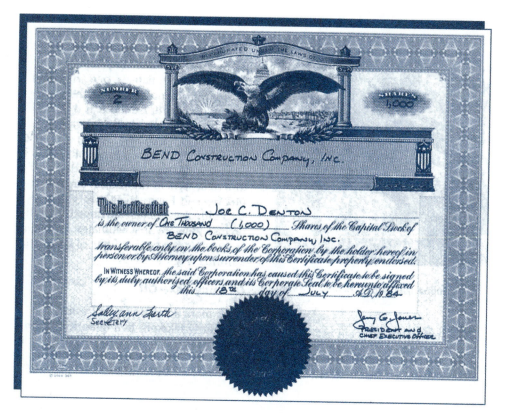

Figure 1–8 Owners of corporations
have shares of stock in the
corporation.

The owners buy shares of stock (Fig. 1–8). A share of stock is a share or a part of the business. The value of each share increases or decreases according to the success of the company. The stocks of many large corporations are bought and sold (traded) in public stock exchanges. Anybody can buy one or more shares of publicly traded stock and be a part owner of that business. Most small corporations and many large corporations are privately held. A privately held corporation is one in which stock is owned only by a select group of investors. Privately held stock cannot be bought and sold through public stock exchanges.

A corporation is managed by its board of directors (Fig. 1–9). The stockholders appoint the board of directors at an annual meeting of the stockholders. In some small corporations, all the owners are on the board of directors. The directors meet regularly to decide the policies and major operating procedures of the company. Managing the day-to-day operation of the company is the responsibility of the president, who is named by the directors.

In a corporation, no person has unlimited liability. The owners can only lose the amount of money they invested in stock. The owners of a corporation are not responsible for the debts of the corporation. The corporation itself is the legal body and is responsible for its own debts. This protection against personal liability is one of the greatest advantages of a corporation. Of course, each person is personally responsible for obeying the law. The shield of a corporation cannot protect a dishonest person who breaks the law in an effort to falsely control the finances of even a large corporation.

Because there is no person who can be held accountable for the actions of the company, the government has stricter regulations for corporations than for the other forms of ownership. Also, corporations are more expensive to form and to operate than are proprietorships and partnerships.

Building Codes

Most towns, cities, and counties have building codes. A building code is a set of regulations (usually in the form of a book) that ensure that all buildings in that jurisdiction (area covered by a certain government agency) are of safe construction. Building codes specify such things as minimum size and spacing of lumber for wall framing, steepness of stairs, and fire rating of critical components. The local building department enforces the local building codes. States usually have their own building codes, and state codes often require local building codes to be at least as strict as the state code. Most small cities and counties adopt the state code as their own, meaning that the state building code is the one enforced by the local building department.

Until recently there were three major **model codes** that were published by independent organizations. (A model code is a suggested building code that is intended to be adopted as is or with revisions to become a government's official code.)

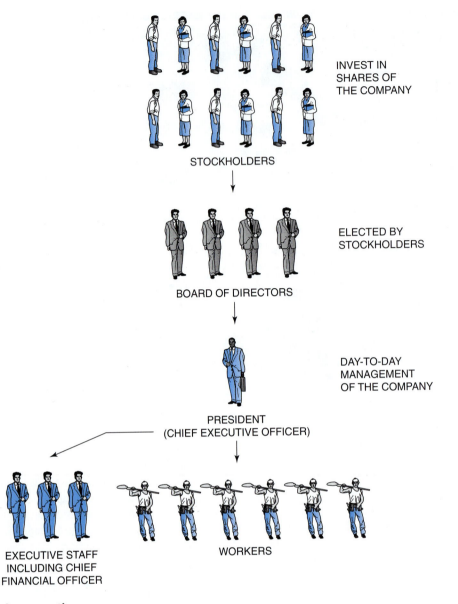

INVEST IN
SHARES OF
THE COMPANY

STOCKHOLDERS

ELECTED BY
STOCKHOLDERS

BOARD OF DIRECTORS

DAY-TO-DAY
MANAGEMENT
OF THE COMPANY

PRESIDENT
(CHIEF EXECUTIVE OFFICER)

EXECUTIVE STAFF
INCLUDING CHIEF
FINANCIAL OFFICER

WORKERS

Figure 1–9 **Structure of a corporation.**

Each model code was widely used in a different region of the United States. By themselves model codes have no authority. They are simply a model that a government agency can choose to adopt as their own or modify as they see fit. In 2003 the International Code Council published a new model code called the *International Building Code* (Fig. 1–10). They also published the *International Residential Code* to cover home construction. Since publication of the first *International Building Code,* states have increasingly adopted it as their building code.

Other than the building code, there are many codes that govern the safe construction of buildings. There are plumbing codes, fire protection codes, and electrical codes. Most

workers on the job site do not need to refer to the codes much during construction. It is the architects and engineers who design the buildings that usually see that the code requirements are covered by their designs. Plumbers and electricians do, however, need to refer to their respective codes frequently. Especially in residential construction, it is common for the plans to indicate where fixtures and outlets are to be located, but the plumbers and electricians must calculate loads and plan their work so it meets the requirements of their codes. The electrical and plumbing codes are updated frequently, so the workers in those trades spend a certain amount of their time just learning what is new in their codes.

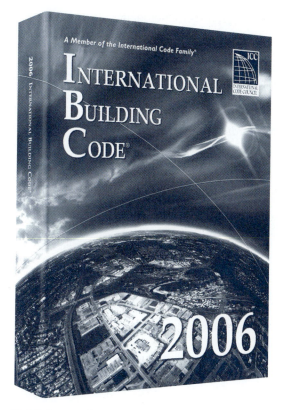

Figure 1–10 *International Building Code,* **courtesy of the International Code Council.**

International Association of Bridge, Structural, Ornamental and Reinforcing Iron Workers (http://www.ironworkers.org/)

International Association of Heat and Frost Insulators and Asbestos Workers (http://www.insulators.org/)

International Brotherhood of Boilermakers, Iron Ship Builders, Blacksmiths, Forgers and Helpers (http://www.boilermakers.org/)

International Brotherhood of Electrical Workers (http://www.ibew.org/)

International Brotherhood of Teamsters (http://www.teamster.org/)

International Union of Bricklayers and Allied Craftworkers (http://www.bacweb.org/)

International Union of Elevator Constructors (http://www.iuec.org/)

International Union of Operating Engineers (http://www.iuoe.org/)

International Union of Painters and Allied Trades (http://www.iupat.org/)

Laborers' International Union of North America (http://www.liuna.org/)

Operative Plasterers' and Cement Masons' International Association of the United States and Canada (http://www.opcmia.org/)

Sheet Metal Workers' International Association (http://www.smwia.org/)

United Association of Journeymen and Apprentices of the Plumbing and Pipefitting Industry of the United States and Canada (http://www.ua.org/)

United Brotherhood of Carpenters and Joiners of America (http://www.carpenters.org/)

United Union of Roofers, Waterproofers and Allied Workers (http://www.unionroofers.com/)

Utility Workers Union of America (http://www.uwua.org/)

Figure 1–11 **Construction craft unions.**

Unions and Contractors' Associations

In the construction industry, there are thousands of organizations of people with common interests and goals. Whole directories of these organizations are available in libraries and on the Internet. Two categories of construction organizations are of particular importance to construction students: craft unions and contractors' associations.

Unions

A **craft union,** usually just called a "union," is an organization of workers in a particular building trade. Workers' unions were first formed in the 1800s when factory workers were being forced to work extreme hours under unsafe conditions—and for very low wages. Although working conditions in both factories and construction have improved dramatically, unions continue to serve a valuable role in the construction industry. Figure 1–11 lists several national construction craft unions.

Union members pay dues to be members of the union. Dues money pays for the benefits the union provides for its members. Most unions have an apprenticeship program that includes both classroom instruction and on-the-job supervised work experience. Some of the members' dues pay for instructors, classroom space, and training supplies. Unions usually provide a pension for members who have worked in the trade. Because they represent a large block of members, unions can be a powerful force in influencing government to do such things as pass worker safety laws, encourage more construction, and support technology that is good for construction. Unions negotiate with employers (contractors) to establish both a pay rate and working conditions for their members. It is quite typical to find that union members enjoy a higher hourly pay rate than nonunion workers in the same trade.

Contractors' Associations

There are associations of contractors that include just about every imaginable type of construction contractor. Figure 1–12 lists only a small number of the largest associations

Air Conditioning Contractors of America
(http://www.acca.org)

Air Conditioning and Refrigeration Institute
(http://www.ari.org)

Associated Builders and Contractors (http://www.abc.org)

National Association of Home Builders
(http://www.nahb.org)

Home Builder's Institute (http://www.hbi.org)

Independent Electrical Contractors Association
(http://www.ieci.org)

National Electrical Contractors Association
(http://www.necanet.org)

National Utility Contractors Association
(http://www.nuca.com)

Plumbing-Heating-Cooling Contractors Association
(http://www.phccweb.org)

The Associated General Contractors (AGC) of America
(http://www.agc.org)

Figure 1–12 **These are only a few of the largest construction associations.**

that have apprenticeship programs. Some contractors' associations are formed to represent only nonunion contractors; a few represent only union contractors; and others represent both. Many associations of nonunion contractors were originally formed because the contractor members felt a need to work together to provide some of the benefits that union contractors receive—such as apprentice training and a lobbying voice in Washington, D.C.

Review Questions

1 Briefly describe the four levels of construction industry workers and give an example of each.

2 What are two ways of getting the training and developing the skills necessary to work in a building trade?

3 Where might you get the knowledge necessary to work as a surveyor?

4 Describe one job that might be done by an engineer in the home building industry.

5 What construction occupation is most apt to be concerned with the arrangement of rooms and the flow of traffic in a house?

6 Describe the relationship between general contractors and subcontractors.

7 Whose seal must appear on the drawings before work can begin on a house?

8 Briefly describe each of the following forms of business ownership:

Sole proprietorship

Limited liability partnership

Corporation

9 In which form of ownership is the risk the greatest?

10 What is the advantage of a corporation to the business owner?

11 Which of the following might be covered by a building code?

a. minimum size of floor framing materials

b. type of building allowed in a certain area

c. distance from road or street to front of building

d. all of the above

12 Which trade does work that is commonly covered by a code?

a. electricians

b. carpenters

c. plumbers

d. all of the above

13 List three things a union does for its members.

14 List three places one might go for training in a building trade.

Activities

CONSTRUCTION PLANNING AND DEVELOPMENT

This activity will help you examine the roles of the various professionals in the planning and construction of a building. Through this activity you will also make valuable contacts with people in the industry you have chosen to enter.

Materials

- Notebook and pencil

PROCEDURE

Find one construction project, preferably a house or housing development under construction. For this activity you will identify the people involved and record their names, the name of their business, and the phone number.

Check with your instructor before you start this activity. Your instructor might want you to work on this in small groups, so that the workers at a construction project are not interrupted several times by members of your class.

Note: Courtesy and respect for others is always helpful. You will be asking several people to give you a couple of minutes of their time. They will be more willing to do that if you follow these rules:

- *Be polite.*
- *Speak up. No one wants to struggle to hear what you are saying.*
- *If the person does not have time to talk right then, ask for an appointment at a more convenient time. Make sure it is a time when you can show up on time.*

- *Know what you are going to say or ask and be organized.*
- *Obey all posted rules and regulations.*
- *Dress neatly and do not wear extreme clothing.*
- *Wear all personal protective equipment required on that site.*

For this activity you note contact information for each of the following:

1. Address and brief description of the construction project
2. Name and address of the architect (you can probably get this from the drawings that the superintendent and foremen should have)
3. Property owner
4. General contractor
5. Superintendent (if the general contractor has appointed one)
6. Concrete or foundation contractor
7. Carpentry (framing) contractor
8. One additional subcontractor
9. Foreman for one building trade (name the trade, specify the company the foreman works for, and include the foreman's name and business phone number)
10. Building inspector (you can get this name from the building department of the city or town where the project is located)

Save this information in an address book. Frequently you will want to contact someone in a particular construction field, and the information you collected in this exercise will be the start of a list of contacts for your future.

Chapter 2

Working in the Industry

After completing this chapter, the student should be able to:

- ✪ discuss ethical issues in the workplace.
- ✪ work as a participating member of a work team.
- ✪ communicate effectively with fellow workers.
- ✪ explain the value of lifelong learning.

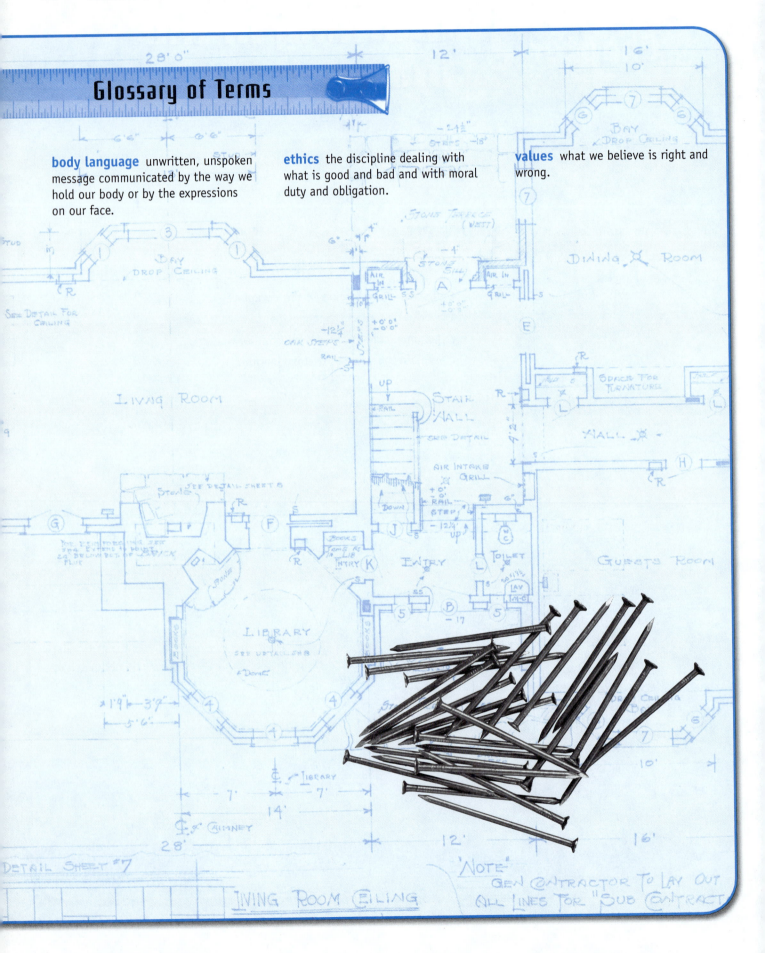

Glossary of Terms

body language unwritten, unspoken message communicated by the way we hold our body or by the expressions on our face.

ethics the discipline dealing with what is good and bad and with moral duty and obligation.

values what we believe is right and wrong.

Often success in a career depends more on how people act or how they present themselves to the world than it does on how skilled they are at their job. Most employers would prefer to have a person with modest skills but a great work ethic than a person with great skills but a weak ethic. In this chapter we will examine some of the non-skill issues that will be most important to you in your career.

Ethics

Ethics are principles of conduct that determine which behaviors are right and wrong. There are two aspects of ethics: values and actions. **Values** have to do with what we believe to be right or wrong. We can have a very strong sense of values, knowing the difference between right and wrong, but not act on those values. If we know what is right but we act otherwise, we lack ethics. To be ethical, we must have good values and act accordingly.

We often hear that someone has a great *work ethic*. That simply means that the person has good ethics in matters pertaining to work. Work ethic is the quality of putting your full effort into your job and striving to do the best job you can. A person with a strong work ethic has the qualities listed in Figure 2–1. Good work ethics become habits, and the easiest way to develop good work ethics is to consciously practice them.

Common Rationalizations

We judge ourselves by our best intentions and our best actions. Others judge us by our last worst act. Conscientious people who want to do their jobs well often fail to consider their behavior at work. They tend to compartmentalize ethics into two parts: private and occupational. As a result, sometimes good people think it is okay to do things at work that they know would be wrong outside of work. They forget that everyone's first job is to be a good person. People can easily fall prey to rationalizations when they are trying to support a good cause. "It is all for a good cause" is an attractive rationale that changes how we see deception, concealment, conflicts of interest, favoritism, and violations of established rules and procedures. In making tough decisions, do not be distracted by rationalizations.

There are great benefits to having good work ethics. As little children, most of us learned the difference between right and wrong. As adults, when we do what we know is right, we feel good about ourselves and what we are doing. On the other hand, doing what we know is wrong is depressing. Although we might think we are getting something for nothing, in that part of us that was programmed as a little child, we know that we have done wrong. We lose respect for ourselves, knowing that what we have done is not something we would want others to do to us. If we make it a habit to do what we know in our hearts not to be right, we develop a general feeling about life and our job. The days seem to go very slowly, and we are happy less often. But if we develop a habit of always trying to do our best, we know that we are doing what is right. Life seems fun, and we look forward to what will come next. Employers recognize people with a good work ethic. They are the people who are always doing something productive, their work turns out better, and they seem cheerful most of the time. Which person do you think an employer will give the most opportunities to: a person who is always busy and whose work is usually well done or a person who seems glum and must always be told what to do next?

Working on a Team

Constructing a building is not a job for one person acting alone (Fig. 2–2). The work at the site requires cooperative effort by carpenters, masons, plumbers, painters, electricians, and others. There are usually several workers from each of these trades. Can you imagine a football game in which each player tries to do it all, without involving his teammates? There would be no blocking, or if there were, it would be in the wrong place. If a pass were thrown, who would catch it? There would be chaos on the field. A construction project without teamwork would have the same kinds of problems. One carpenter's work would not match up with another carpenter's work. There would be too much of some materials and not enough of others. Walls would be enclosed before the electricians ran the wiring in them.

Teamwork is very important on a construction site, but what does being a team player on a construction team mean? Effective team members have the best interests of the whole team at heart. Each team member has to carry his or her own load, but it goes beyond that. Sometimes a team member might have to carry more than his or her own load, just because that is what is best for the team. If you are installing electrical boxes and the plumber says one of your boxes is in the way of a pipe, it might be in the best interests of the project to move the electrical box. That would mean you would have to undo work you had just completed and then redo it. It is, after all, a lot easier to relocate an outlet box than to reroute a sink drain.

A person with a strong work ethic:
• shows up to work a few minutes early instead of a few minutes late.
• looks for a job to do as soon as the previous one is done. (This person is sometimes described as a self-starter.)
• does every job as well as possible.
• stays with a task until it is completely finished.
• looks for opportunities to learn more about the job.
• cooperates with others on the job.
• is honest with the employer's materials, time, and resources.

Figure 2–1 **Characteristics of good work ethic.**

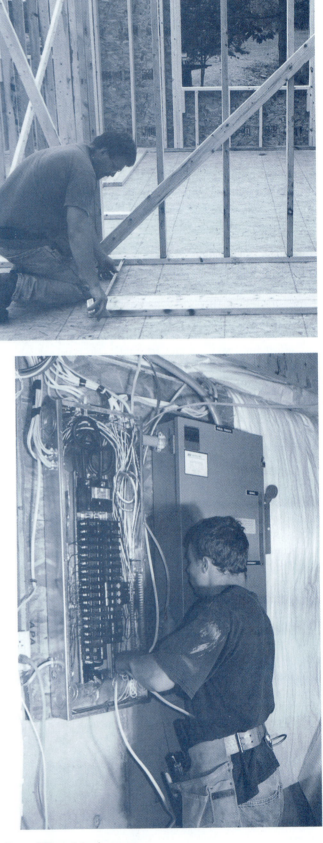

Figure 2–2 Work on the job requires cooperative effort by individuals from different trade areas.

The following are six traits of an effective team:

- *Listening*. Team members listen to one another's ideas. They build on teammates' ideas.
- *Questioning*. Team members ask one another sincere questions.
- *Respect*. Team members respect one another's opinions. They encourage and support the ideas of others.
- *Helping*. Team members help one another.
- *Sharing*. Team members offer ideas to one another and tell one another what they have learned.
- *Participation*. Team members contribute ideas, discuss them, and play an active role together in projects.

Communication

Remember that football team. How could members function as a team without communication? Good communication is one of the most important skills for success in any career. Employers want workers who can communicate effectively; but more importantly, you must be able to communicate with others to do your job well and to be a good team member. Look back at the six traits of an effective team in the last section and ask yourself which traits require communication?

There are many forms of communication, but the most basic ones are speaking, listening, writing, reading, and body language. If you master those five forms of communication, you will probably succeed in your career.

Speaking

To communicate well through speech, you need a reasonably good vocabulary. It is not necessary, or even desirable, to fill your speech with a lot of flowery words that do not say much or that you do not really understand. What is necessary is to know the words that convey what you want the listener to hear, and it is equally necessary to use good enough grammar so those words can be communicated properly. Using the wrong word or using it improperly can cause two serious problems: For one thing, if you use the wrong word, you will not be saying what you intended to say. This is also often true if you use a great word wrong since you still might not be saying what you thought you were saying. For another thing (the second serious problem), using a poor choice of words or using bad grammar gives the listener the impression that you are poorly educated or that maybe you just do not care about good communication skills. As a businessperson, you will find that communicating is critical to earning respect as a professional as well as to gaining people's business. Look your listeners in the eye. Ask yourself if you think they understand what you are saying. If it is important, ask them if they understand. If they do not understand, try a different approach. The best way to develop good speaking skills is by practicing them—even when you are just with your friends. And your friends will be much more impressed with your effort to speak well than they would be with your ability to speak poorly.

Listening

Good listening is an important skill. Have you ever had people say something to you, and after they were finished and gone, you wondered what they said or you missed some of the details? Perhaps they were giving you directions or telling you about a school assignment. If only you could listen to them again! If possible, try paraphrasing. Paraphrasing means to repeat what they said but in different words. If someone gives you directions, wait until the person is finished. Then repeat the directions to person, so he or she can tell you if you are correct. Look at the speaker and form a mental picture of what the speaker is saying. Make what the speaker is saying important to you. Good listening can mean hearing and acting on a detail of a job that will result in giving a competitive edge in bidding.

Writing

Writing is a lot like speaking, except you do not have the advantage of seeing if the person seems to understand or of asking if the person understands. That means you really have to consider your reader. If you are giving instructions, keep them as simple as possible. If you are reporting something to a supervisor, make your report complete, but do not take up his or her time with unrelated trivia. Penmanship, spelling, and grammar count. Always use good grammar to ensure that you are saying what you intend and that your reader will take you seriously. Use standard penmanship, and make it as neat as possible. Do not invent new ways of forming letters, and do not try to make your penmanship ornate. You will only make it harder to read. If you are unsure of how to spell a word, look it up in a dictionary. Next time, you will know the word and will not have to look it up. After you write something—read it, thinking about how your intended reader will take it. Make changes if necessary. Your writing is important! Sole proprietors have to demonstrate good writing skills in proposals and contracts. If either of these is poorly written, it can cost the business a lot of money.

Reading

You will have to read at work. That is a fact no matter what your occupation. You will have to read building specifications, instructions for use of materials and tools, safety notices, and notes from the boss. To develop reading skills, find something you are interested in and spend at least 10 or 15 minutes every day reading it. You might read the sports section of the newspaper, books about your hobby, hunting and fishing magazines, or anything else that is interesting to you. What is important is that you read. Practicing reading will make you a better reader. It will also make you a better writer and a better speaker. When you come across a word you do not know how to pronounce or you do not know the meaning of, look it up or ask someone for help. You will find that you learn pronunciation and meaning very quickly, and your communication skills will

Figure 2–3 Body language is an important form of communication.

improve faster than you expect. In practically no time, you will not need help very often.

Body Language

Body language is an important form of communication. How you position your body and what you do with your hands, face, and eyes all convey a lot of information to the person you are communicating with (Fig. 2–3). Whole books are written about how body language is used to communicate and how to read body language. We will only discuss a couple of key points here.

When you look happy and confident, the message you convey is that you are honest (you have nothing to hide or to worry about) and you probably know what you are talking about. If you look unhappy, unsure of yourself, or uninterested, your body language tells the other person to be wary of what you are saying—something is wrong. The following are a few rules for body language that will help you convey a favorable message:

- Look the other person in the eye. Looking toward the floor makes you look untrustworthy. Looking off in space makes you seem uninterested in the other person.
- Keep your hands out of your pockets, and do not wring your hands. Just let your hands rest at your sides or in your lap if you are sitting. An occasional hand gesture is okay, but do not overdo it.
- Dress neatly. Even if you are wearing work clothes, you can be neat. Faddish clothes, extra baggy or extra tight fitting clothes, and T-shirts with offensive messages on them all distract from the real you.
- Speak up. How loudly you speak might not seem like body language, but it has a lot to do with how people react to you. If they have to strain to hear what you are saying, they will think that either you are not confident in what you are saying or you are angry and not to be

trusted. If you see your listeners straining to hear you or if they frequently ask you to repeat what you are saying, speak a little louder.

Customer Service

In any industry, you will only be as successful as you are good at building your reputation for doing quality work and for the degree to which your customers are happy with you and your job. On the job site, your customer might be a crew chief, a foreman, a subcontractor, or a contractor. If you are the contractor or subcontractor, the customer will be whoever hired you. It doesn't actually matter who hired you, though—your role will always be to do the very best job you can for whomever it is that you are working.

Good customer service also includes providing a good value for your fees, being honest, communicating clearly, being cooperative, and looking to provide the best possible experience your customer can have in working with you. Just as when you practice good ethics, when you provide great customer service you will enjoy your job much more. You will be proud of your work, others will want to hire you more often, and your career will be much easier to build. Think about how you like being treated when you are a customer—and always try to treat your customers at least as well.

Lifelong Learning

Lifelong learning refers to the idea that we all need to continue to learn throughout our entire lives. Not so long ago, those who were fortunate enough to get a formal education started it in elementary school; and if they did well and their families did not need them to help with work and support, they went to high school. In their teens, most men went to work or started an apprenticeship, and most women either worked in low-paying jobs or did housework. Only the wealthy and very fortunate went on to college. An apprentice worked alongside a skilled craftsman and learned a skilled trade. Apprentices were generally indentured, meaning that in return for learning a trade with a journeyman, they were committed to working for that journeyman for a specified number of years. Those great craftspeople of the past usually spent their entire working lives in the same job.

We have greater opportunities to learn and greater opportunities to move up a career ladder today. Our lives are filled with technology, innovative new materials, and new opportunities. People change not only jobs, but entire careers several times during their working life. Those workers who do not understand the new technology in the workplace, along with those who do not keep up with the changes in how their company is managed, are destined to fall behind—not stay even, but actually fall behind economically. There is little room in a fast-paced company of this century for a person whose knowledge and skills are not growing as fast as the company. If all you know are the techniques, equipment, and materials that were current 10 or 15 years ago, and if you have not made any attempt to learn the skills necessary to move up in the management of the company, you will not be nearly as valuable to the company as you were when those skills were the state of the art.

To keep up with new information and to develop new skills for the changing workplace, everyone must continue to learn throughout life. Some high school graduates still choose the time-tested route of entering college immediately after high school graduation. Others enlist in the military or begin an apprenticeship. All these graduates are continuing to learn so that they will be better prepared to work in the modern, changing world. Are they finished learning after college or the military or when they complete their apprenticeship? No, the worker is no more finished with the need to learn than the world is finished changing. Companies today send their employees to special classes that pertain to their jobs or to possible future jobs. Many adult workers, often people in their 40s and 50s, fill college classrooms. It is not uncommon for an engineer or a doctor to attend classes at a community college with the intention of starting a new career. The organizations that train large numbers of apprentices (unions, trade associations, very large employers, etc.) frequently list more courses for journeymen than they do for apprentices. That's because those journeymen recognize the importance of continuing to learn throughout their careers. Throughout this textbook there are profiles of construction workers, most of whom have continued their education by formal or informal studies during their careers.

Review Questions

1. Who has the primary responsibility for making decisions about your education and training for work?

2. Which is likely to have the greatest impact on your success in the career you have chosen: use of proper English, your attitude, or your skill in the work you do?

3. How would you define "ethics"?

4. What benefits are there for doing what is ethically right?

5. Explain one reason why people suffer from their own unethical behavior.

6. Describe the characteristics of a good team member.

7. List five forms of communication.

8. How can you know if people understand you when you are speaking to them?

9. How can you verify that you understand what someone else is saying to you?

10. List four things that will help you communicate better in writing.

11. Describe three things a construction worker would have to read. Describe situations that require a paragraph or more of reading.

12. Describe how a person could use body language to communicate that they are interested in what someone else is saying to them.

13. Describe what the term "lifelong learning" means.

14. Why has lifelong learning become increasingly important in the construction industry?

Activities

OCCUPATIONAL WORK ETHIC INVENTORY

This activity is reproduced here with the permission of its author, Gregory C. Petty, Ph.D., University of Tennessee.

For each work ethic descriptor listed below, select the answer that most accurately describes your standards for that item. There are no right or wrong answers. You will not be graded on this activity. It is simply a good way to get to know yourself, so respond to each item as honestly as possible. There is no time limit, but you should work as rapidly as possible. Please respond to every item on the list. If you have access to the Internet, you may do this activity online at http://www.coe.uga.edu/cgi-bin/cgiwrap/~rhill/new_owei/owei.pl. That way, it can be automatically scored. Check with your teacher about whether to do it online or from this textbook.

As a worker I can describe myself as:

1. Dependable
   - ☐ never
   - ☐ almost never
   - ☐ seldom
   - ☐ sometimes
   - ☐ usually
   - ☐ almost always
   - ☐ always
2. Stubborn
   - ☐ never
   - ☐ almost never
   - ☐ seldom
   - ☐ sometimes
   - ☐ usually
   - ☐ almost always
   - ☐ always
3. Following regulations
   - ☐ never
   - ☐ almost never
   - ☐ seldom
   - ☐ sometimes
   - ☐ usually
   - ☐ almost always
   - ☐ always
4. Following directions
   - ☐ never
   - ☐ almost never
   - ☐ seldom
   - ☐ sometimes
   - ☐ usually
   - ☐ almost always
   - ☐ always
5. Independent
   - ☐ never
   - ☐ almost never
   - ☐ seldom
   - ☐ sometimes
   - ☐ usually
   - ☐ almost always
   - ☐ always
6. Ambitious
   - ☐ never
   - ☐ almost never
   - ☐ seldom
   - ☐ sometimes
   - ☐ usually
   - ☐ almost always
   - ☐ always
7. Effective
   - ☐ never
   - ☐ almost never
   - ☐ seldom
   - ☐ sometimes
   - ☐ usually
   - ☐ almost always
   - ☐ always
8. Reliable
   - ☐ never
   - ☐ almost never
   - ☐ seldom
   - ☐ sometimes
   - ☐ usually
   - ☐ almost always
   - ☐ always

9. Tardy
   - ☐ never
   - ☐ almost never
   - ☐ seldom
   - ☐ sometimes
   - ☐ usually
   - ☐ almost always
   - ☐ always

10. Initiating
    - ☐ never
    - ☐ almost never
    - ☐ seldom
    - ☐ sometimes
    - ☐ usually
    - ☐ almost always
    - ☐ always

11. Perceptive
    - ☐ never
    - ☐ almost never
    - ☐ seldom
    - ☐ sometimes
    - ☐ usually
    - ☐ almost always
    - ☐ always

12. Honest
    - ☐ never
    - ☐ almost never
    - ☐ seldom
    - ☐ sometimes
    - ☐ usually
    - ☐ almost always
    - ☐ always

13. Irresponsible
    - ☐ never
    - ☐ almost never
    - ☐ seldom
    - ☐ sometimes
    - ☐ usually
    - ☐ almost always
    - ☐ always

14. Efficient
    - ☐ never
    - ☐ almost never
    - ☐ seldom
    - ☐ sometimes
    - ☐ usually
    - ☐ almost always
    - ☐ always

15. Adaptable
    - ☐ never
    - ☐ almost never
    - ☐ seldom
    - ☐ sometimes
    - ☐ usually
    - ☐ almost always
    - ☐ always

16. Careful
    - ☐ never
    - ☐ almost never
    - ☐ seldom
    - ☐ sometimes
    - ☐ usually
    - ☐ almost always
    - ☐ always

17. Appreciative
    - ☐ never
    - ☐ almost never
    - ☐ seldom
    - ☐ sometimes
    - ☐ usually
    - ☐ almost always
    - ☐ always

18. Accurate
    - ☐ never
    - ☐ almost never
    - ☐ seldom
    - ☐ sometimes
    - ☐ usually
    - ☐ almost always
    - ☐ always

19. Emotionally stable
    - ☐ never
    - ☐ almost never
    - ☐ seldom
    - ☐ sometimes
    - ☐ usually
    - ☐ almost always
    - ☐ always
20. Conscientious
    - ☐ never
    - ☐ almost never
    - ☐ seldom
    - ☐ sometimes
    - ☐ usually
    - ☐ almost always
    - ☐ always
21. Depressed
    - ☐ never
    - ☐ almost never
    - ☐ seldom
    - ☐ sometimes
    - ☐ usually
    - ☐ almost always
    - ☐ always
22. Patient
    - ☐ never
    - ☐ almost never
    - ☐ seldom
    - ☐ sometimes
    - ☐ usually
    - ☐ almost always
    - ☐ always
23. Punctual
    - ☐ never
    - ☐ almost never
    - ☐ seldom
    - ☐ sometimes
    - ☐ usually
    - ☐ almost always
    - ☐ always
24. Devious
    - ☐ never
    - ☐ almost never
    - ☐ seldom
    - ☐ sometimes
    - ☐ usually
    - ☐ almost always
    - ☐ always
25. Selfish
    - ☐ never
    - ☐ almost never
    - ☐ seldom
    - ☐ sometimes
    - ☐ usually
    - ☐ almost always
    - ☐ always
26. Negligent
    - ☐ never
    - ☐ almost never
    - ☐ seldom
    - ☐ sometimes
    - ☐ usually
    - ☐ almost always
    - ☐ always
27. Persevering
    - ☐ never
    - ☐ almost never
    - ☐ seldom
    - ☐ sometimes
    - ☐ usually
    - ☐ almost always
    - ☐ always
28. Likeable
    - ☐ never
    - ☐ almost never
    - ☐ seldom
    - ☐ sometimes
    - ☐ usually
    - ☐ almost always
    - ☐ always

29. Helpful
   - ☐ never
   - ☐ almost never
   - ☐ seldom
   - ☐ sometimes
   - ☐ usually
   - ☐ almost always
   - ☐ always

30. Apathetic
   - ☐ never
   - ☐ almost never
   - ☐ seldom
   - ☐ sometimes
   - ☐ usually
   - ☐ almost always
   - ☐ always

31. Pleasant
   - ☐ never
   - ☐ almost never
   - ☐ seldom
   - ☐ sometimes
   - ☐ usually
   - ☐ almost always
   - ☐ always

32. Cooperative
   - ☐ never
   - ☐ almost never
   - ☐ seldom
   - ☐ sometimes
   - ☐ usually
   - ☐ almost always
   - ☐ always

33. Hard working
   - ☐ never
   - ☐ almost never
   - ☐ seldom
   - ☐ sometimes
   - ☐ usually
   - ☐ almost always
   - ☐ always

34. Rude
   - ☐ never
   - ☐ almost never
   - ☐ seldom
   - ☐ sometimes
   - ☐ usually
   - ☐ almost always
   - ☐ always

35. Orderly
   - ☐ never
   - ☐ almost never
   - ☐ seldom
   - ☐ sometimes
   - ☐ usually
   - ☐ almost always
   - ☐ always

36. Enthusiastic
   - ☐ never
   - ☐ almost never
   - ☐ seldom
   - ☐ sometimes
   - ☐ usually
   - ☐ almost always
   - ☐ always

37. Cheerful
   - ☐ never
   - ☐ almost never
   - ☐ seldom
   - ☐ sometimes
   - ☐ usually
   - ☐ almost always
   - ☐ always

38. Persistent
   - ☐ never
   - ☐ almost never
   - ☐ seldom
   - ☐ sometimes
   - ☐ usually
   - ☐ almost always
   - ☐ always

39. Hostile
- ☐ never
- ☐ almost never
- ☐ seldom
- ☐ sometimes
- ☐ usually
- ☐ almost always
- ☐ always

40. Dedicated
- ☐ never
- ☐ almost never
- ☐ seldom
- ☐ sometimes
- ☐ usually
- ☐ almost always
- ☐ always

41. Devoted
- ☐ never
- ☐ almost never
- ☐ seldom
- ☐ sometimes
- ☐ usually
- ☐ almost always
- ☐ always

42. Courteous
- ☐ never
- ☐ almost never
- ☐ seldom
- ☐ sometimes
- ☐ usually
- ☐ almost always
- ☐ always

43. Considerate
- ☐ never
- ☐ almost never
- ☐ seldom
- ☐ sometimes
- ☐ usually
- ☐ almost always
- ☐ always

44. Careless
- ☐ never
- ☐ almost never
- ☐ seldom
- ☐ sometimes
- ☐ usually
- ☐ almost always
- ☐ always

45. Productive
- ☐ never
- ☐ almost never
- ☐ seldom
- ☐ sometimes
- ☐ usually
- ☐ almost always
- ☐ always

46. Well groomed
- ☐ never
- ☐ almost never
- ☐ seldom
- ☐ sometimes
- ☐ usually
- ☐ almost always
- ☐ always

47. Friendly
- ☐ never
- ☐ almost never
- ☐ seldom
- ☐ sometimes
- ☐ usually
- ☐ almost always
- ☐ always

48. Loyal
- ☐ never
- ☐ almost never
- ☐ seldom
- ☐ sometimes
- ☐ usually
- ☐ almost always
- ☐ always

49. Resourceful
    ☐ never
    ☐ almost never
    ☐ seldom
    ☐ sometimes
    ☐ usually
    ☐ almost always
    ☐ always
50. Modest
    ☐ never
    ☐ almost never
    ☐ seldom
    ☐ sometimes
    ☐ usually
    ☐ almost always
    ☐ always

To score the Occupational Work Ethic Inventory yourself, see Appendix A on page 276.

ETHICS

Most behavior on the work site is governed by rules, regulations, and common sense. Ph.D. Christopher Bauer of Bauer Ethics Seminars in Nashville, Tennessee, works with companies who want to build and maintain great ethics. He points out that in some situations, it isn't at all clear what the right thing to do. When there are multiple options for solving a problem, but each requires some type of ethical compromise, that is called an ethical dilemma. Here is an example: You discover that a worker at your site got his job by lying about his work experience. You know that many deserving and more qualified workers ought to have that job and that, in fact, this individual may not be completely safe doing certain required tasks because of his lack of experience. However, he has done an outstanding job so far, is well liked, and the job will likely fall behind if time is taken to replace him.

   What would you do in this case?
   Which of your values support informing your supervisor about what you have found out?
   Which of your values support saying nothing?

Do you find yourself using rationalizations to support your decision? If so, what kind of discomfort are your rationalizations trying to reduce?
If you were to confront this worker, what would you say?

TEAMWORK

You have been selected to be on a team of five specialists who will fly by helicopter into a remote location in Alaska to remodel a lodge that is to be used for a meeting of several world leaders, including the President of the United States. The five of you have been equipped with the finest tools available, and you will be dropped off at this wilderness location for four weeks to convert the lodge into a place that can properly host presidents and prime ministers.

You encounter unexpected bad weather, and your helicopter is forced to land in a forested area 30 miles southeast of your destination and more than 33 miles southwest of the nearest town. Fortunately, no one was seriously injured in the rough landing, but one of your construction crew did receive a nasty sprained ankle and some bruised ribs. The radio in your helicopter was damaged in the landing, and so you have no way of contacting anyone else and there is no one else at the lodge yet. That means it will be at least three days before anyone knows you are stuck in the wilderness. The date is October 14. Daytime temperatures are in the 50s, but it drops to well below freezing at night, and on the day of your forced landing, there is a cold rain and it is windy.

The following are the items you have with you in the helicopter. You do not have anything other than the helicopter itself and the items on this list.

   circular saw
   cordless drill
   reciprocating saw
   propane torch
   2 pipe wrenches
   3 tape measures
   assorted screwdrivers

electrical pliers

vise grip

4-foot level

plumb bob with 10 feet of string

chalk line with 100 feet of string

electrical tape

110 lb. of assorted nails

50 lb. of assorted screws and bolts

electric meter

4 dozen pencils

12 candy bars

5 one-quart bottles of water

compass

navigation charts

sewing kit

personal toiletries (toothbrush, shaving supplies, etc.) for each team member

two changes of clothes and a light jacket for each team member

What should you do? If you decide to hike to the lodge or to town, you will have to contend with your teammate's sprained ankle, and you will have to decide what to take with you. If you decide to stay and wait for help, you will have to figure out how to keep warm at night and how to find food.

Ponder this scenario on your own and decide what you think should be done and what equipment you should use, and for what purpose. When you have a plan that you think is the best solution, form a team of five with your classmates and develop a team plan. Review the six traits of an effective team in this chapter before starting work. Do your best to be an effective team member. How did your individual plan compare with the team plan? The team plan will almost always be the best plan.

Safety

SECTION TWO
SAFETY

Success Stories

Donna Story

TITLE

President and founder, U.S. Plumbing in Charlotte, Tennessee

EDUCATION

Donna completed high school and holds an associate's degree in computer accounting. She has also completed the plumber's state licensing exam in Tennessee.

HISTORY

Donna quickly realized that a desk job in computer accounting was not for her. She took her first plumbing job with H & H Plumbing in Nashville. In 2001, she started a new business with her husband, Steve.

ON THE JOB

Donna and Steve work about 10 hours a day, starting with an hour in the office each morning to set up the day's work. They divide the site work between them. Donna reports: "I run copper for water lines and do the soldering. Steve runs plastic for drain lines." They complete billing and other paperwork at night. Donna hires and manages a few high school students during the summer.

BEST ASPECTS

"It's always rewarding to see the finished product," Donna reflects. "I would never have dreamed that so much work goes into plumbing a house. When homeowners are proud and pleased, that's a big reward."

CHALLENGES

Donna describes the physical challenge of working in crawl spaces: "Residential work is a little more challenging than commercial because you're often lying on your back all day under a house" in cramped conditions. She has also worked hard to build confidence in a male-dominated field. "A lot of women out there don't like office jobs," Donna notes. "They're outside people. But they're so intimidated by the construction field. I'm not easily intimidated."

IMPORTANCE OF EDUCATION

"It takes a lot of math skill to read blueprints," states Donna, who values her success in high school math courses. College courses in psychology have helped her learn to work with people. "You're often working directly with the customer," she says, reminding that "on a remodel, you're working in their living quarters. You must have good people skills."

FUTURE OPPORTUNITIES

Within 5 to 10 years, Donna wants to employ and supervise about five people. She hopes to plumb grocery stores one day, and will consider earning a general contractor's license.

WORDS OF ADVICE

"Plumbing is a wonderful field with a good living to be made. See if you like it. If you do, then go for the gusto. The sky's the limit."

Chapter 3 | Jobsite Safety

OBJECTIVES

After completing this chapter, the student should be able to:

- explain what an accident is and what causes accidents.
- define OSHA and explain its impact on construction workers.
- identify the appropriate PPE for common work situations and explain how to use it.
- explain the Hazard Communication Standard and find information on a Material Safety Data Sheet.
- explain how fires are ignited, sustained, and extinguished.
- recognize the dangers of working in trenches and explain how to work safely in a trench.
- explain electric shock and list safety considerations in working around electricity.

Glossary of Terms

ampere the unit of measurement of electric current.

class A fire a fire that involves ordinary materials like paper, cardboard, and wood. Class A fires can be extinguished with water.

class B fire a fire that involves flammable liquids. Class B fires are extinguished with either dry chemicals or CO_2 (carbon dioxide).

class C fire an electrical fire. Class C fires are extinguished with CO_2 (carbon dioxide).

conductor a material that allows electricity to flow.

fire triangle consists of heat, fuel, and oxygen (the three sides of the triangle); the three elements must be present for a fire to burn.

ground (electrical) a conducting body that serves as the common return path for an electric circuit. A ground typically has zero potential. The earth may also be used as a ground.

ground fault circuit interrupter (GFCI) a protective device that opens the electric circuit when an imbalance in the amount of current flow between the conductors is sensed.

horseplay practical jokes and playful activity that are inappropriate on a construction site.

labels user instructions found on most construction products. The product label contains valuable information.

Material Safety Data Sheet (MSDS) gives complete information about the product and what to do in the event of exposure. An MSDS is required to be available for any substance that might be harmful.

OSHA refers both to the state and federal Occupational Safety and Health Administration and also to state administrations. OSHA also stands for the Occupational Safety and Health Act, which is administered by the Occupational Safety and Health Administration. OSHA generally refers to the laws that are intended to keep workers safe.

personal protective equipment (PPE) any safety equipment you wear to protect yourself from safety hazards.

Right to Know rule the OSHA rule that says that every worker has a right to know about any substances on the job that might be harmful to humans.

voltage the electrical pressure that causes current to flow, measured in volts (sometimes abbreviated as V). Voltage is sometimes called electromotive force (EMF) because it is the force that causes electrons to move. Voltage also represents the difference of potential, or potential difference, in a circuit.

working conditions the things in the work environment that affect your work.

work practices the things a worker does and how he or she works—these practices have a lot to do with safety.

afety on a construction job site is critically important. Construction can be dangerous work if the workers are uninformed about how to be safe or if their attitude is that safety is less important than other aspects of the job. Workplace safety is so important that the federal government has special laws just to ensure that workplaces are kept safe and that workers have the knowledge they need to work safely. Safety is every worker's responsibility. The best of laws and the most safety-conscious employer cannot protect a worker who does not know safe practices or does not have a good safety attitude.

Accidents

The National Safety Council says that nearly 4 million people suffer a disabling injury at work each year and 100,000 people are killed in workplace accidents annually. Accidents are avoidable. They are caused by either poor work practices or poor working conditions.

Work Practices

Work practices refers to how we work. Poor work practices cause many workplace accidents. The following are all examples of poor work practices:

- **Performing tasks without proper training**
Many tasks involved in construction can be dangerous if done incorrectly. We often work with sharp cutting tools, power equipment, or heavy loads. When untrained workers who want to feel mature and want to show how capable they are try to do these jobs, they endanger themselves and others around them. For example, shoveling concrete seems like a job anyone can do. However, many people do not know that fresh concrete can cause severe chemical burns. Also, excessive handling is not good for the concrete. As a student, you will be learning the proper way to perform many tasks. Make sure you have learned the proper way before attempting any task.

- **Failing to read labels and instructions**
Most of the products we use in construction come with labels or user instructions. The labels are found on products that include everything from paint and adhesive to windows, doors, and power tools. The labels and instructions tell the user what hazards might be involved in the use of the product and how to avoid them (Fig. 3–1). Reading a label is not a substitute for proper instruction on how to use a tool or perform an operation, but it can give you information about that specific product or tool that general training did not give. For example, the label will tell you if a particular adhesive is highly flammable and should only be used in a

Figure 3–1 Product labels often contain safety information.

well-ventilated area. Imagine what could happen if you use this adhesive in a small enclosed room with a doorbell and someone rings the bell. (Doorbells can cause an electric spark.)

- **Ignoring safety devices**
Using safety devices seems like a pretty obvious safety rule, but it is one that is often violated. Safety devices such as guards on circular saws are there because workers were injured before the safety device was invented. You might feel like the guard is getting in your way or obstructing your vision until you become accustomed to working with the guard in place, but experienced carpenters can work just as quickly and accurately with all the guards in place as they can with the guards removed. You might think you can be safe by paying attention and not putting your hands near an unguarded saw blade or unprotected belt drive. When you use a tool several times every day, day after day, you become very familiar with it. The day is bound to come when you are in a hurry and are not paying as close attention to that unguarded equipment. The injury that results can be for life—all because you did not think you needed a guard. The same lesson applies to electrical protection, fire protection, and all other safety devices.

- **Engaging in horseplay**
Horseplay refers to all sorts of practical jokes and playful activity that are inappropriate on a construction site. As a member of a construction crew, you will become friends with the other members of the team. It is good to enjoy the company of the people you work with, but horseplay can be dangerous. Practical jokes distract attention from a potentially dangerous job. Be friendly, but be professional.

- **Having poor communication**
There are times when communication on the job site is critical. Many accidents are caused by people not communicating well. Good communication requires complete and accurate messages, careful listening, and a clear understanding of who is responsible for what information or messages. Consider what could happen if one electrician is working on a circuit and tells a helper to go to another part of the house and disconnect (turn off the circuit breaker at) circuit 19; then another worker tells the helper "no, he's working on circuit 17."

- **Using alcohol and drugs**
This may seem obvious, but many accidents are caused by alcohol or drug use. Even some prescription drugs can slow your reaction time, make you drowsy, or affect your judgment. Would you hold a nail while a person who has been drinking alcohol swings the hammer? A construction site is no place for a person who has been using alcohol or drugs, and the effect of these substances can linger in your body for hours. If you take prescription drugs, ask your doctor how they will affect you at work. Many employers require drug and alcohol testing of all their employees.

Figure 3–2 **A messy site can be a safety hazard.**

Working Conditions

Working conditions are the things in the work environment that affect your work. Poor working conditions may be caused by the workers on the site, or they may be a result of nature and the location of the site. The following are some poor working conditions:

- **Poor housekeeping**
Housekeeping is important on a construction site. Construction activity generates a lot of scrap and debris (Fig. 3–2). Accidents are bound to happen when the workers are stumbling over scraps beneath their feet and stepping over tools and materials. At the end of each day and as each step of your work is completed, clean up the site. It only takes a few minutes to pick up your scrap and sweep the area, and those few minutes will make the site safer and more pleasant.

- **Wet conditions**
Usually it is not necessary to work in puddles of water. If the site is reasonably well graded, water will not collect in puddles. Still there will be times when the grading is not done and rain will collect in puddles. If you have to work where there is a puddle, fill it with soil from nearby high ground, drain it, or bridge it with sound planks. Never work with power tools or electricity around water.

- **Excessive heat or cold**
There is not much you can do to change the weather, but you can adjust your work practices according to the weather. If it is extremely hot, dress accordingly, wear sunscreen, and drink plenty of fluids. If it is extremely cold, wear layers of warm clothing. Remind yourself that either of these conditions can reduce your alertness, cause you to be less nimble, and cause accidents. Work cautiously.

- **Poor lighting**
It is difficult to do quality work when you cannot see the details, and it can be dangerous. If you are working after dark or in a poorly lighted area, take the time to

obtain lighting. If your work will move around, you might need several lights so that you do not have to constantly move a single work lamp, but do not be tempted to work without adequate light.

- **Poorly maintained or inoperative equipment**
It is every worker's responsibility to keep track of the condition of the equipment. Construction work is demanding, and equipment will need to be repaired or replaced from time to time. Never use power equipment, ladders, scaffolding, or tools that are not in good working condition. Poorly maintained equipment is one of the most common causes of accidents. If a piece of equipment you are using needs attention, tell your foreman or supervisor.

- **Fire and explosion hazards**
A number of things can present a danger of fire or explosion—*never* work around them. (How fires start and how they continue to burn are discussed later in this chapter.) There is never a reason why you will need to work in an explosive atmosphere or in a place where there is real danger of fire. If the danger is an explosive atmosphere, eliminate the source and ventilate the area before working. If you are working near a potential source of fuel for a fire, do not do work that can ignite that fuel. Spontaneous combustion is a fire that starts itself. Know what causes spontaneous combustion and avoid it. Know where the nearest fire extinguisher is and how to use it.

OSHA

What Is OSHA?

Every worker has a right to a safe and healthful workplace. That is why Congress passed the Occupational Safety and Health Act of 1970, requiring employers to provide workplaces free from serious recognized hazards and to comply with occupational safety and health standards. The Occupational Safety and Health Administration (OSHA) wants every worker to go home whole and healthy every day. The agency was created by Congress to help protect workers by setting and enforcing workplace safety and health standards and by providing safety and health information, training, and assistance to workers and employers. If you work in the private sector, you are covered by an OSHA regional office under federal OSHA or an OSHA program operated by your state government.

The Occupational Safety and Health Act authorizes states to establish their own safety and health programs with OSHA approval. Twenty-three states operate state OSHA programs covering private-sector workers as well as state and local government employees. (In addition, Connecticut, New York, and New Jersey cover state and local government employees only.) State OSHA programs must be at least as effective as the federal program and provide similar protections for workers. Some states set their own standards; others adopt federal rules. All state programs conduct inspections and respond to worker complaints.

The Occupational Safety and Health Act grants workers important rights. Workers have a vital role to play in identifying and correcting problems in their workplaces, working with their employers whenever possible. Often, employers will promptly correct hazardous conditions called to their attention. But workers also can complain to OSHA about workplace conditions threatening their health or safety.

What Are Employees' Responsibilities?

OSHA requires workers to comply with all safety and health standards that apply to their actions on the job. Employees should:

- read the OSHA poster that every employer is required to display (Fig. 3–3).
- follow the employer's safety and health rules and wear or use all required gear and equipment.
- follow safe work practices for their job, as directed by their employer.
- report hazardous conditions to a supervisor or safety committee.
- report hazardous conditions to OSHA if employers do not fix them.

What Are Employers' Responsibilities?

The Occupational Safety and Health Act requires employers to provide a safe and healthful workplace free of recognized hazards and to follow OSHA standards. Employers' responsibilities also include providing training, medical examinations, and record keeping.

Throughout the OSHA standards there are requirements for the employer to provide a "competent person" to build certain safety devices or to inspect safety protection of workers. OSHA defines "competent person" in 29 CFR 1926.32(f) as "one who is capable of identifying existing and predictable hazards in the surroundings or working conditions which are unsanitary, hazardous, or dangerous to employees, and who has authorization to take prompt corrective measures to eliminate them."

What Is an OSHA Standard?

OSHA issues standards or rules to protect workers against many hazards on the job. These standards limit the amount of hazardous chemicals that workers can be exposed to, require the use of certain safety practices and equipment, and require employers to monitor hazards and maintain records of workplace injuries and illnesses. Employers can be cited and fined if they do not comply with OSHA standards.

You Have a Right to a Safe and Healthful Workplace.

IT'S THE LAW!

- You have the right to notify your employer or OSHA about workplace hazards. You may ask OSHA to keep your name confidential.

- You have the right to request an OSHA inspection if you believe that there are unsafe and unhealthful conditions in your workplace. You or your representative may participate in the inspection.

- You can file a complaint with OSHA within 30 days of discrimination by your employer for making safety and health complaints or for exercising your rights under the *OSH Act*.

- You have a right to see OSHA citations issued to your employer. Your employer must post the citations at or near the place of the alleged violation.

- Your employer must correct workplace hazards by the date indicated on the citation and must certify that these hazards have been reduced or eliminated.

- You have the right to copies of your medical records or records of your exposure to toxic and harmful substances or conditions.

- Your employer must post this notice in your workplace

The *Occupational Safety and Health Act of 1970 (OSH Act)*, P.L. 91-596, assures safe and healthful working conditions for working men and women throughout the Nation. The Occupational Safety and Health Administration, in the U.S. Department of Labor, has the primary responsibility for administering the *OSH Act*. The rights listed here may vary depending on the particular circumstances. To file a complaint, report an emergency, or seek OSHA advice, assistance, or products, call 1-800-321-OSHA or your nearest OSHA office: • Atlanta (404) 562-2300 • Boston (617) 565-9860 • Chicago (312) 353-2220 • Dallas (214) 767-4731 • Denver (303) 844-1600 • Kansas City (816) 426-5861 • New York (212) 337-2378 • Philadelphia (215) 861-4900 • San Francisco (415) 975-4310 • Seattle (206) 553-5930. Teletypewriter (TTY) number is 1-877-889-5627. To file a complaint online or obtain more information on OSHA federal and state programs, visit OSHA's website at **www.osha.gov**. If your workplace is in a state operating under an OSHA-approved plan, your employer must post the required state equivalent of this poster.

1-800-321-OSHA
www.osha.gov

U.S. Department of Labor • **Occupational Safety and Health Administration** • **OSHA 3165**

Figure 3–3 This poster is to be displayed on every job.

Personal Protective Equipment

Personal protective equipment (PPE) is any safety equipment you wear to protect yourself from safety hazards. It may be more general, such as a hard hat or safety glasses, which protect you from a variety of hazards; or it may be intended for a specific purpose. For example, a safety harness is intended to prevent you from falling from a high place. Your safety depends on your PPE—that is why it is vital to keep it clean and in good condition. This textbook cannot cover all the possible requirements for PPE, and so you will need to follow the regulations of your employer in compliance with OSHA. The following are a few common items of PPE:

- Safety shoes (Fig. 3–4) are available with hard toes to protect against being crushed by heavy objects from above. Not all occupations require hard-toe shoes, but most construction occupations require them at least part of the time. Puncture-proof shoes protect against punctures from nails and other sharp objects. Puncture-proof shoes have hard soles. All shoes worn on a construction site should be puncture-proof.
- Hard hats protect against injury from falling objects. Hard hats are always required when anyone is working above you. Many construction sites require all workers to wear hard hats at all times. Hard hats have an outer shell and an inner webbing that can be adjusted to fit your head size. Hard hats are to be made of a material that will not conduct electricity (Fig. 3–5).

- Safety glasses protect against small particles injuring your eyes. Approved safety glasses are stamped Z87.1 and have side shields (Fig. 3–6). Safety glasses should be worn when there is even a slight chance of small particles flying into your eyes. This includes sawdust; chips from hammering masonry, concrete, or hardened nails; metal particles from grinding; and so on. There are many situations that require more complete eye

Figure 3–5 Hard hat. *Courtesy of Northern Safety Co. Inc., Frankfort, NY.*

Figure 3–4 Safety shoes. *Courtesy of Northern Safety Co. Inc., Frankfort, NY.*

Figure 3–6 Safety glasses. *Courtesy of Northern Safety Co. Inc., Frankfort, NY.*

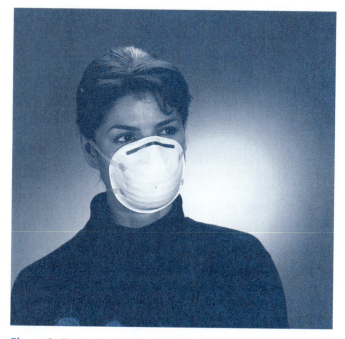

Figure 3–7 **Particulate mask.** *Courtesy of Northern Safety Co. Inc., Frankfort, NY.*

Figure 3–8 **Safety harness.** *Courtesy of Northern Safety Co. Inc., Frankfort, NY.*

protection or protection from bright light. For example, welders wear special eye protection that filters out the harmful light of welding.

- A particulate mask is usually a disposable mask worn over the nose and mouth to protect you from airborne particles of dust (Fig. 3–7). A particulate mask is worn when sanding drywall joints, doing extensive wood sanding, and cutting concrete. A particulate mask is not a gas mask and will not remove any harmful gas from the air.

- A safety harness (Fig. 3–8) should be worn whenever you work where it would be possible to fall more than 6 feet. (A guardrail system may be used to protect workers from falling. Guardrails must be built by someone who has received the necessary training and is certified as a "competent person" according to OSHA.) The safety harness is attached to a secure anchor point in such a way that you are free to move about and work, but if you fall, the harness will stop your fall within 6 feet. The first few times you wear a safety harness, you might feel awkward, but a properly fitted harness is not uncomfortable and will not restrict your movements. You will get used to it quickly.

- Ear protection is required when working around loud equipment. Approximately 1 out of every 10 Americans has suffered some hearing loss, and exposure to loud noise for a prolonged period of time is the main cause. Sound does not have to be painful to cause hearing damage. According to the U.S. Environmental Protection Agency, continued exposure to noise at

70 decibels (approximately the amount of noise from a hair dryer) can begin to damage your hearing. The OSHA rules for what type of hearing protection is required under various conditions are fairly complex, taking both noise level and length of time into account. The best way to protect your hearing on the job site is to wear hearing protection whenever you work around loud noises. Hearing protection can be as simple as foam earplugs, which will block most of the damaging sound from most construction operations. Another type of hearing protection is external ear muffs (Fig. 3–9).

Hazard Communication Standard

Many of the materials used in construction can cause sickness or even death. Some materials cause nerve damage, some cause damage to organ tissue, some cause skin irritation (sometimes very severe), and some cause respiratory or circulatory system damage. Other materials may be hazardous because of the risk of fire or explosion. There are too many potential threats to health and safety to list them all here or for you to learn them all in this course. Instead of trying to memorize all the hazardous materials used in construction and their harmful effects, you should know that there are many and that the only way to be safe is to read and obey warning labels and follow safety precautions.

Figure 3–9 **Hearing protection.** *Courtesy of Northern Safety Co. Inc., Frankfort, NY.*

CAUTION

CAUTION: Never dispose of hazardous materials (hazmats) or other chemicals by pouring them into a drain or on the ground.

OSHA has a rule that every worker has a right to know about any substances on the job that might be harmful to humans. This rule is the *Hazard Communication Standard—* often called the HazCom Standard or the **Right to Know rule.** It might not seem that construction would involve many hazardous chemicals, but there are very many substances found on construction sites that fall under the Right to Know rule. A few of the most common are paint, sawdust, concrete, concrete dust, adhesives, solvents, plaster dust, and many types of insulation.

According to the HazCom Standard or Right to Know rule, every worker who comes in contact with any potentially dangerous material must have access to a **Material Safety Data Sheet (MSDS).** The MSDS describes what the risks are, how to guard against them, and what to do in case of a hazmat exposure (Fig. 3–10). MSDSs are not generally packaged with building products, but they are readily available from the manufacturer and can usually be found on the manufacturer's web site.

A few simple rules can eliminate most of the danger from common hazardous materials.

• Always keep chemicals and solvents in their original containers, with the manufacturer's label in place. It is very dangerous to use chemicals without knowing what they are.

• Read labels. Before you use even the most ordinary supplies, read the manufacturer's label completely. If there is any danger, the manufacturer usually explains what the danger is, how to avoid it, and what to do if an accident occurs.

• Read the MSDS for each hazardous or questionable product with which you come in contact. The MSDS will give you a lot more information than the product label.

• Use chemicals and other supplies only for their intended purpose. There is a solvent, adhesive, coating, or other chemical designed for just about anything you might want to do. Why try to force another substance to do the job?

• Follow all recommended safety precautions, and use the recommended PPE for every product and operation.

Fire

The Fire Triangle

In chemical terms, fire is the rapid oxidation of fuel. What this means is that when some fuel rapidly combines with oxygen, the result is fire. When iron rusts, it is combining with oxygen, or oxidizing; but the process is slow, and so it is not fire. Another characteristic of fire is that it produces heat and usually light. Rusting does not produce heat or light. Three elements must be present for a fire to start: energy to ignite the fire (usually in the form of heat), fuel, and oxygen. These three elements together form the **fire triangle** (Fig. 3–11). Let's take a closer look at the three parts of the fire triangle.

Heat

Heat raises the fuel to its flash point, the temperature at which it ignites. Heat can come from an open flame, from a smoldering or hot substance near the fuel, from an electric spark, or from a chemical reaction. Spontaneous combustion occurs when a chemical reaction takes place in a confined space, where no ventilation is available to remove the heat the chemical reaction gives off. The best-known example of this is when oily or solvent-soaked rags are piled together. The chemicals in the oil or solvent react with the fabric in the rags, generating a large amount of heat. Because no airflow can get into the pile of rags, the heat builds up and eventually reaches the flash point of the rags. They might just smolder until the high temperature reaches the surface of the pile or someone disturbs the pile. Then flames burst out.

Fuel

Fuel for a fire can be anything that is capable of being oxidized rapidly enough to give off heat. Common fuels around a construction site are wood, paper, rags, plastics, solvents, paint, and electrical insulation. Of course, there are many other possibilities.

Kleenz #402
Material Safety Data Sheet

Section I – Identity

Polk Incorporated
4114 Sand Road
Albany, NY 12002

Telephone number for information
(518) 555-1212

Date prepared
06/05/2000

Hazardous - Yes

Health	1
Flammability	2
Reactivity	0
Personal Protection	B

4 = Most Hazardous
Emergency Phone Number
(800) 555-1234 or Poison Control Center

Section II – Hazardous Ingredients

Hazardous Components CAS Reg#	PEL ppm	TWA ppm	STEL	Other Limits	% <opt>
Stoddard solvent CAS# 8052–41–3	500	100			<50

LISTED CHEMICAL SUBJECT TO REPORTING REQUIREMENTS OF SEC.313 TITLE III

Section III – Physical/Chemical Characteristics

Boiling Point	350°F	Specific Gravity (Water = 1)	0.87
Vapor Pressure (mm Hg)	1.5	Melting Point	NA
Vapor Density (Air =1)	4.5	Evaporation Rate (Butyl Acetate = 1)	0.08
Viscosity (CPS)	NA	Decomposition Temperature	NA
Auto Ignition Temp.	NA	Solubility in Water	Negligible
PH	NA		

Appearance and Odor – clear amber liquid, solvent odor

Section IV – Fire and Explosion Hazard Data

Flash Point – TCC 140F

Flammable Limits
LEL 0.7   UEL–6.0

Extinguishing Media – CO_2, Foam, Dry Chemical

Fire fighting procedure – cool containers with water, use self-contained breathing apparatus, BURNING PRODUCT IS TOXIC.

Fire and Explosion Hazards – CLOSED CONTAINERS MAY BURST WHEN EXPOSED TO EXTREME HEAT. ISOLATE FROM HEAT, SPARKS, AND OPEN FLAME.

Section V – Reactivity Data

Stability – Stable
Conditions to Avoid – Heat, sparks, open flame.
Incompatibility – Strong oxidizing agents.
Hazardous Decomposition or Byproduct – Thermal decomposition may yield carbon monoxide.
Hazardous Polymerization – Will not occur.
Conditions to Avoid – NA.

Section VI—Health Hazard Data

Routes of Entry:
Inhalation—Yes   Skin—No   Ingestion—Yes
Health Hazards—Acute: Causes eye irritation, drying of skin, dizziness.
          Chronic: Respiratory tract irritation, dermatitis.
          Target Organs: Skin, eyes, respiratory system.

Figure 3–10 Material Safety Data Sheet.

Carcinogenity: No
NTP–NA   IARC Monographs–NA   OSHA Regulated–Yes
Signs and Symptoms of Exposure–Irritation and redness of skin, dizziness
Medical Conditions Aggravated–Dermatitis
                              Broncho–Pulmonary Problems

Emergency First Aid Procedures:
Eye and Skin Contact: Blot with towel, flush with large quantities of water.
Inhalation: Remove to fresh air.
Ingestion: Induce vomiting, get immediate medical attention.

Section VII–Precautions for Safe Handling and Use

Steps to be taken in case the material is released or spilled–Dike and collect material into metal container. USE NON-
    SPARKING TOOLS. Use water or foam spray to lessen fire hazard. Use absorbent for small spill.

Waste Disposal Method–Recycle, incinerate, or use hazardous waste management facility for disposal according to state and
    federal regulations. EPA Waste No. D001

Precautions to be taken in Handling and Storage–Combustible. Keep away from heat, sparks, and open flame.Use adequate
    ventilation. Avoid prolonged or repeated contact.

Other Precautions–Store in secure area. Dispose of empty containers safely. Keep out of reach of children.

Section VIII–Control Measures

Respiratory Protection–Use self-contained breathing apparatus for concentrations above threshold limit value.

Ventilation   Local Exhaust–Recommended.
               Mechanical–Required; exchange every 10–15 minutes.
               Special–Blow air into pits and depressions.
               Other–Assure fresh air circulation.

Protective Gloves–Required; rubber, plastic.
Eye Protection–Required; safety glasses.
Other Protection–Apron or shop coat recommended.

Work/Hygienic Practices–Do not eat, drink, or smoke next to material. Wash hands and face after work.

The information herein is based on data considered accurate. However, no warranty is expressed or implied regarding the
accuracy of these data or results to be obtained from the use thereof.

Figure 3–10 Continued.

Figure 3–11 Fire triangle.

Oxygen

The air is the most common source of oxygen for accidental fires, but it is not the only source. Magnesium is a
metal that burns under water with an extremely hot flame.
The oxygen for an underwater magnesium fire comes from
the water. Some welding processes use pure oxygen, supplied

in tanks. When pure oxygen is present, the danger of fire is
increased greatly.

The Fire Tetrahedron

It was once thought that the elements of the fire triangle
were the only ones involved in a fire, but fire scientists have
added a fourth element, a chain reaction. The four elements
of fire make up the fire tetrahedron (Fig. 3–12). (A tetrahedron is a solid shape with four faces, like a pyramid.) The
chain reaction occurs when the fuel is broken down by the
heat of the fire, making the fuel easier to oxidize.

Extinguishing Fires

Everyone should have some basic knowledge of how to extinguish fires. If any of the three parts of the fire triangle is
missing, the fire will not ignite. Without energy, the chain

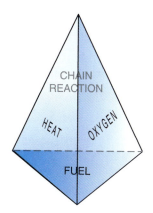

Figure 3–12 **Fire tetrahedron.**

reaction cannot be started. Without oxygen, oxidation cannot occur. Without fuel, there is nothing to oxidize (burn). If the fourth part of the tetrahedron, the chain reaction, is missing, the fire will not continue. Generally, the heat generated by burning the fuel is enough to keep the chain reaction going. To prevent accidental fires, all we have to do is keep heat or oxygen away from fuel. Preventing fires is always better than having to extinguish them. To extinguish (stop) a fire, all we have to do is remove one of the elements of the fire tetrahedron.

Not all fire extinguishers will work on all types of fires. There are four classes of fires, and there are fire extinguishers designed to extinguish each class.

- **Class A fires** involve ordinary materials like paper, cardboard, and wood. Class A fires can be extinguished with water, which cools and prevents the oxygen in the air from reaching the fuel. Class A fires can also be extinguished with some dry chemicals, which blanket the surface and block oxygen from reaching the fuel. Not all dry chemical extinguishers can be used on class A fires.
- **Class B fires** involve flammable liquids like paint, solvents, and gasoline. Class B fires are extinguished with either dry chemicals, which blanket the surface and remove oxygen, or CO_2 (carbon dioxide), which cools the fire and displaces oxygen. Dry ice is carbon dioxide, and so it is a good cooling agent. Never use a water extinguisher on a class B fire, because many flammable liquids are lighter than water and the water will only spread the fire.
- **Class C fires** are electrical fires. Before an electrical fire is extinguished, the electrical supply should be disconnected, to prevent the fire from reigniting. CO_2 extinguishers are used for class C fires. Never use water on an electrical fire because of the risk of electric shock.
- Class D fires are those involving combustible metals like magnesium. Class D fires are less common in construction because we do not normally work with combustible metals. However, if those metals are being used, a class D fire extinguisher should be on hand.

Figure 3–13 **Fire extinguisher labels show the types of fires they can be used for.** *Courtesy of Kidde Safety Products.*

Many fire extinguishers are rated for more than one class of fire. The label on the extinguisher will tell you what classes of fires that extinguisher can be used for (Fig. 3–13). Do not work without a suitable fire extinguisher on the site.

Trench Safety

Working in a trench can be very dangerous. Each year hundreds of workers are killed in trench cave-ins. OSHA defines a trench as an excavation that is deeper than it is wide and is less than 15 feet wide at the bottom. Trenches are common on construction sites. They are necessary to bury electric, water, gas, and sewer lines, and they are sometimes necessary to place building footings.

Normally soil is quite stable, with surrounding soil causing equal pressure on the soil beside it. However, when a trench is excavated, there is nothing exerting pressure on the side of the excavation. The pressure of the soil that remains on the unsupported side of the excavation can cause the side to collapse. Vibrations from machinery such as a backhoe can contribute to the collapse. Depending on the type of soil, it can weigh between 2,000 and 3,000 pounds per cubic yard. A person buried under 2 feet of soil will be under enough weight to prevent him or her from breathing. Death can result in minutes. A trench cave-in might release several yards of soil weighing many tons.

OSHA has specific rules for trench safety. Those rules are beyond what can be covered in this book, but they include things like how much slope the walls of a trench must have to make it safe and what types of shoring (mechanical support) can be used to make trench work safe. To be safe, do not work in a trench unless someone who has been trained as a "competent person" according to OSHA requirements for trench work agrees that the trench is safe.

Material Handling

Construction materials are often heavy, bulky, or long. If they are handled improperly they can present a safety hazard. Always lift heavy objects with your legs, not your back, as shown in Figure 3–14. Bend your knees into a squatting position, keeping your upper body upright. Then grasp the load and straighten your legs to lift it. Your back is not designed to carry a heavy load in a bent-over position.

If you cannot lift the object easily by yourself because it is too heavy or too bulky, have a partner help you. Carrying a bulky load that blocks your vision as you walk through a congested work site is an invitation for an accident. Be particularly aware when you carry long objects like lumber, ladders, and pipe. While you are watching the front of the load, the back can be hitting people, electrical lines, or ladders.

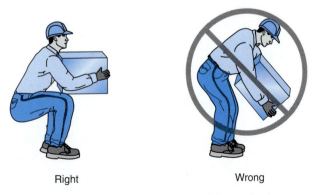

Right   Wrong

Figure 3–14 Lift with your legs, not with your back.

Electricity

To discuss electrical safety, it is necessary to have a basic understanding of electricity. The difficulty in understanding electricity lies in the fact that you cannot see electricity.

Electrical Fundamentals

Perhaps the easiest way to understand electricity is to picture a water fountain, as shown in Figure 3–15. The pump forces water through the pipes, out of the fountain head, and back into the reservoir. The pump picks up the water that has replenished the reservoir, and the cycle continues.

Now look at the electric circuit in Figure 3–16. The battery is the source of electrical pressure, or **voltage.** The battery acts like a pump, supplying the pressure to the circuit. The **conductors,** shown in Figure 3–16, can be compared to pipes. They carry the electric current to and from the lamp. The switch in Figure 3–16 turns the electric circuit on and off by breaking and completing the path for the electric current. If the switch is off, the circuit cannot be completed. This would be like putting a shutoff valve in one of the pipes in the fountain circuit.

Now imagine a hole developed in one of the fountain's pipes (Fig. 3–17). As a result, the water finds an easier path to take and leaks out of the pipe. In turn, there is a loss in pressure, and so the water will not leave the fountain head with as much force as before. Since the pump can only supply a certain amount of pressure, the leak reduces the

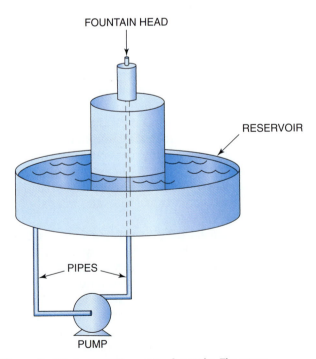

Figure 3–15 An operating water fountain. The pump pressurizes the water, forcing it through the pipes and out of the fountain head to be recirculated.

SWITCH
(OFF)

LAMP
(LOAD)

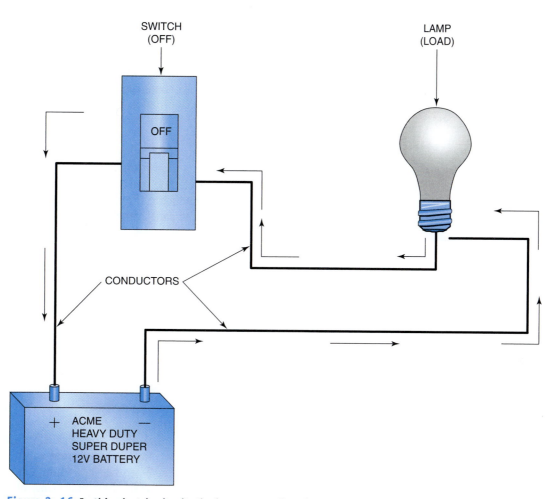

OFF

CONDUCTORS

+ ACME
HEAVY DUTY
SUPER DUPER
12V BATTERY −

Figure 3–16 In this electric circuit, the battery supplies the pressure to force the electric current through the conductors, causing the light to light, and back to the battery. The battery can be compared to the reservoir and pump in the fountain circuit. The conductors (wires) are compared to the pipes.

amount of pressure available to shoot the water from the fountain head.

In Figure 3–18, another conductor has been added to the electric circuit. This conductor provides an alternative path in which the electric current can flow. Just as the water did in the fountain example, electric current will take the path of least resistance. The new conductor provides an easier path for the electric current because there is not a load (light) in this circuit. Some current will still flow through the light—however, not enough to cause the light to light. The majority of the current will take the easier path through the new conductor. The new conductor is called a short circuit.

Electrical Safety

How much electricity does it take to kill someone? This question is not easily answered because there are many factors to consider. The size and weight of the person as well as the person's physical makeup (whether the person is muscular or has a lot of fatty tissue) has an effect on how readily electric current passes through the body. Other factors, such as how well the person is connected to the source of the electricity and how much voltage is present, have an effect as well. Finally, where the electricity enters the body and where it exits the body also impacts the amount of damage caused by the electric current. Figure 3–19 lists different amounts of electric current (measured in **amperes,** and often referred to as amps or A) and their effects on the human body. A general statement to keep in mind is that 0.03 ampere is often enough current to kill a person! The voltage is not what kills; it is the current. However, the voltage is the force or pressure that causes the current to flow. Therefore, a low voltage such as 5 volts may not cause a fatal amount of current to flow, but 100 volts may result in a fatal current flow.

The drill in Figure 3–20 operates on 120 volts and uses approximately 1.3 amperes of electricity when operating. A typical electrical receptacle is rated at 120 volts and 15 amperes. The electrical receptacle contains three contact points for the plug to make an electrical connection (Fig. 3–21). The larger slot of the receptacle will mate with the wider

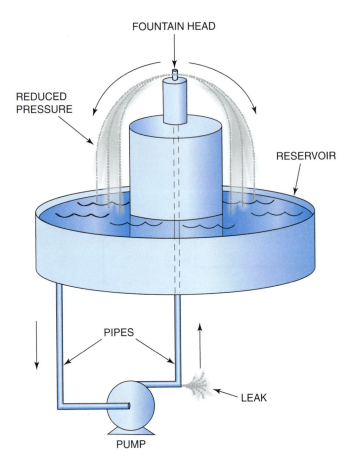

FOUNTAIN HEAD

REDUCED
PRESSURE

RESERVOIR

PIPES

LEAK

PUMP

Figure 3–17 A leak in the pipe causes a loss of pressure.

blade of the plug. This is called the neutral (or grounded) conductor and is identified with white insulation. This conductor is intentionally grounded at the main panel. The smaller slot of the receptacle will mate with the smaller blade of the plug. This is called the hot (or ungrounded) conductor and has black insulation. As the name implies, this conductor is not grounded. Finally, the half-round opening in the receptacle is called the grounding conductor and is identified with green insulation. This conductor is grounded directly to the earth.

We will begin by imagining that the drill is an older model that has a plug with only two blades (hot and neutral). There is no ground prong on the plug and therefore no connection to the ground wire in the receptacle (Fig. 3–22). The drill will work just fine with this arrangement and can be used with relative safety as long as certain failures do not occur within the drill.

Now imagine that some damage has occurred to the wiring leading to the drill plug. The insulation of the hot (ungrounded) conductor has become chafed, causing the exposed conductor inside to come into contact with the metal frame of the drill. The drill continues to operate normally. Even though the hot (ungrounded) conductor is touching the metal frame of the drill, the circuit breaker in the main

electrical panel does not blow or trip. Remember that a normally operating drill uses approximately 1.3 amperes of current. The circuit supplying the drill is rated at 15 amperes. When the hot (ungrounded) conductor comes into contact with the metal frame, the current increases to 10 amperes. This is insufficient current to cause the fuse to blow or the circuit breaker to trip.

But when you grasp the handle in order to use the drill, you receive an electric shock! How did this happen?

You became a path for the electric current (Fig. 3–23). While some electric current is flowing from the hot (ungrounded) conductor, through the drill frame, to the ground and back to the panel, an alternative path now exists. Some of the current will flow from the drill frame through your body, to the ground, and back to the electrical panel. The current flowing through your body may be fatal. Recall that 0.03 ampere is often enough to kill. Yet the additional current is still not enough to cause the circuit breaker to trip!

Now imagine that the drill in our example is equipped with a properly connected three-prong plug. The electric shock would be much less. The third prong of the plug connects the frame of the drill to the electrical **ground** at the panel through the grounding conductor. This is usually called simply the ground.

This time when you grasp the drill, the current again tries to flow through your body to the ground. However, with the three-prong plug, not as much current will flow through your body. Recall that current will follow the path of least resistance. When the hot (ungrounded) conductor comes into contact with the metal frame of the drill, the current finds a more direct or easier path to take. The resistance of the grounding conductor is far less than the resistance of the actual earth. Therefore, most of the electric current takes the easier path and flows through the grounding conductor back to the panel, rather than through you and the earth. Should enough current flow through the grounding conductor, the circuit breaker will trip, removing power from the drill.

Ground Fault Circuit Interrupters

Figure 3–24 shows the same scenario as above, except that you now find yourself standing in a puddle of water when you grasp the drill. The water causes you to be better connected to the ground. This lowers your resistance to ground, which allows more current to flow through your body. The increased current increases the likelihood that the electric shock that you receive may cause injury or even death.

Figure 3–25 shows a device called a **ground fault circuit interrupter (GFCI).** A GFCI is a protective device used to prevent the kind of deadly shock noted above from happening.

In the normally functioning drill circuit, there is 1.3 amperes of current flowing from the electrical panel, through the black wire, through the drill, and back to the panel

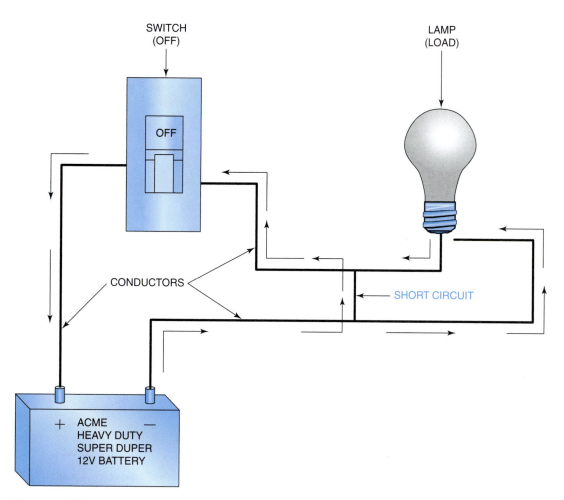

Figure 3–18 A short circuit prevents the light from lighting.

Amount of Current	Effect	Seriousness
0.001 ampere	Slight sensation	None
0.006–0.010 ampere	Muscular contraction	Cannot let go
0.020–0.050 ampere	Asphyxia	Often fatal
0.050–0.200 ampere	Ventricular fibrillation	Probably fatal
0.0250–4.0 ampere	Thoracic muscle constriction (stoppage of heart muscle)	Fatal

Figure 3–19 The effects of electric current on the human body.

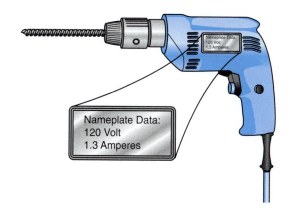

Nameplate Data:
120 Volt
1.3 Amperes

Figure 3–20 Nameplate data from a power tool.

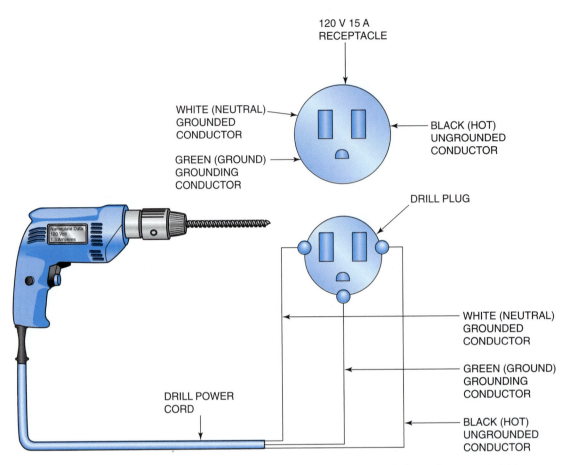

Figure 3–21 A three-conductor power cord contains a hot (black, ungrounded) conductor, a neutral (white, grounded) conductor, and a ground (green or bare, grounding) conductor.

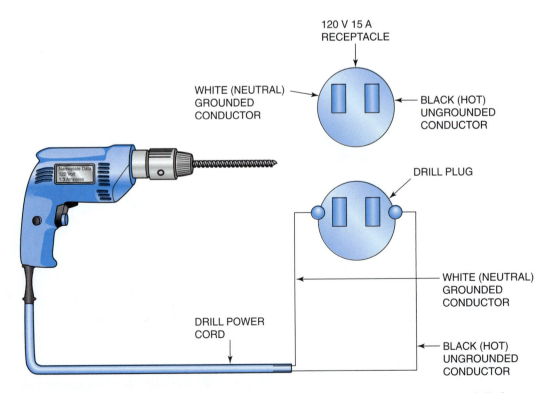

Figure 3–22 Some devices, especially older ones, do not use a three-conductor power cord. Notice the grounding conductor is missing.

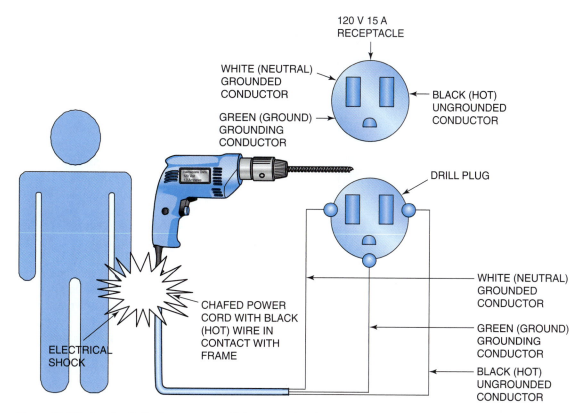

Figure 3–23 A potentially dangerous condition can exist should the hot conductor come into contact with the metal frame of the tool.

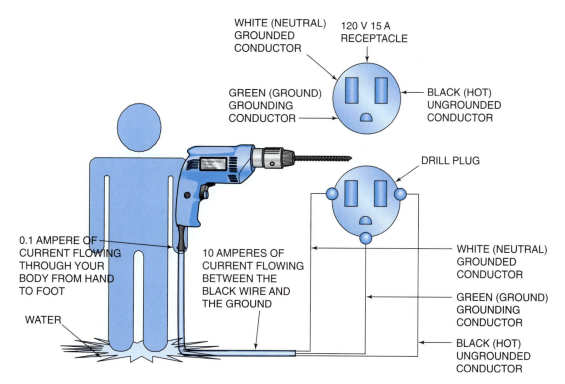

Figure 3–24 Standing in water increases the risk of receiving a fatal shock.

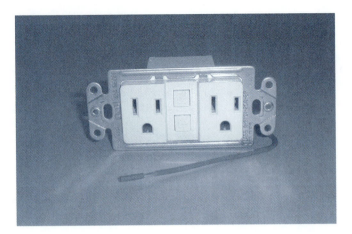

Figure 3-25 A GFCI.

through the white wire. The current that flows through the black wire (1.3 amperes) will be the same value as the current returning on the white wire (1.3 amperes).

A GFCI senses a difference in the amount of current flowing in the black and white conductors. A normally operating circuit will have the same amount of current flowing in each conductor. This is said to be balanced. When a fault occurs (Fig. 3-24), an imbalance in the current between the two conductors will exist. In Figure 3-24, there is 10 amperes of current flowing in the black wire as a result of the chafed insulation and contact with the metal frame of the drill. An additional 0.1 ampere of current is flowing through the black wire, through your body, to ground. Since 10.1 amperes of current is flowing through the black wire, 10.1 amperes of current should return through the white wire. However, this is not the situation. The path for the fault current caused by the chafed wire in the drill cord is between the hot wire (black) and the earth. The path for the additional current caused by your contact with the drill is also between the hot wire (black) and the earth. There will be very little current flowing in the grounded (white) wire. The GFCI senses that the current flow between the black and white conductors is no longer balanced, and this causes the GFCI to trip. When the GFCI trips, power is removed from the drill. GFCIs are designed to trip when they sense as little as 0.006 ampere difference in current between the black and white wires. This is well below the 0.03 ampere that is considered to be often fatal.

Avoiding Electric Shock

The easiest way to avoid an electric shock is to not come into contact with electricity in the first place. Another way to avoid an electric shock is to stay focused on the task at hand and use safe procedures and techniques when working around electricity. Here are steps to take to minimize the risks of working with electricity:

- Know the location of the electrical panel or disconnect before beginning the work.

- Complete a CPR certification course. Encourage your coworkers to do the same.
- Be certain that all your tools, test equipment, power cords, and so on, are in good condition and working order.
- Do not use electric devices in wet or damp locations unless they are specifically designed for this environment.
- Should you need to use a ladder, use a fiberglass model. Wood and aluminum ladders can conduct electricity should they come in contact with a live wire.
- Do not work on an energized circuit.
- Before performing any type of work or act, think about the consequences of that work or act. Do not take chances or cut corners!
- Always work carefully. Place yourself in the safest position to avoid slipping, falling, or backing into energized parts.
- Make certain you are satisfied that you are working in the safest environment possible. Do not continue if you feel uncomfortable due to your own actions or those of a coworker.
- Remove all jewelry (watches, rings, bracelets, etc.). Gold is an excellent conductor of electricity.
- Discard all worn, defective, or otherwise unsafe items.
- De-energize, lock out, and tag out all electric circuits upon which work is to be performed. The purpose of the lockout-tagout procedure is to prevent needless injury or death by removing all hazardous energy from the circuit on which you are working. An electrical lockout involves turning off the circuit breaker that supplies power to the circuit. Then, using a lockout hasp and padlock (Fig. 3-26), lock the handle of the circuit breaker in the off position so that it cannot be turned back on until all lockouts are removed (Fig. 3-27). All workers who are working on the circuit must place their own padlock on the circuit breaker. This is why the lockout hasp has more than one hole in which to place a padlock. Typically, a lockout hasp can accommodate up to six individual padlocks (Fig. 3-28). To inform others who have placed the

Figure 3-26 A lockout hasp, padlock, and key.

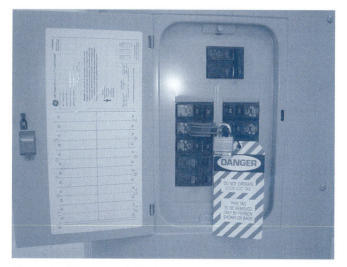

Figure 3–27 **A properly locked out circuit breaker. Notice the lockout tag.**

Figure 3–28 **Three separate padlocks and tags on one lockout hasp for locking out one circuit breaker.**

Figure 3–29 **Front and back sides of a lockout tag.**

padlock, a lockout tag (Fig. 3–29) is used. You must also follow one additional safety step when performing a lockout/tagout. Be certain that no one, other than you and your supervisor or other designated person, has the key to your padlock. After you complete the lockout/tagout, put the padlock key in your pocket. This ensures that no one other than yourself or your supervisor can remove the padlock. This provides additional safety by preventing others from restoring power to the circuit while you are still working on it. A lockout should never be cut off or removed unless both keys are lost and both you and your supervisor are present and in agreement.

• Verify that the circuit is de-energized before beginning your work.
   a. Perform a visual inspection of your meter and test leads.

   b. Determine the amount of voltage that you expect to measure, and verify that the voltage rating of your meter and test leads will not be exceeded.
   c. Set up your meter to measure voltage as follows:
      1. Insert the black test lead into the *COM* test lead jack.
      2. Make certain the test lead is fully and securely inserted into the *COM* jack.
   d. Insert the red test lead into the *volt/ohm* or *V* test lead jack. Make certain the test lead is fully and securely inserted into the *volt/ohm* or *V* jack.
   e. Determine the type of voltage (AC or DC), and set the voltage selection switch (if so equipped) to the proper voltage type.
   f. Determine the approximate amount of voltage to be measured, and set the range switch to the proper voltage range. Again, verify that the voltage rating of your meter and test leads will not be exceeded.
   g. If you are unsure of the amount of voltage that you will be measuring, *set the range switch to the highest voltage range.* If you have an auto-ranging meter, you will not need to set the voltage range. The meter will determine the correct range automatically.
   h. Double-check your meter setup.
   i. Connect the red and black test probes to a circuit with a similar known amount of voltage present.
      1. When measuring voltage, your test leads will be connected across, or in parallel with, the portion of the circuit you are testing.
      2. The safest method for making the meter connections is to de-energize the circuit to be tested, connect your meter into the circuit, and then re-energize the circuit. Unfortunately, in reality, this is impractical. Therefore, exercise extreme caution when connecting your meter to an energized circuit. Be sure to wear appropriate PPE, as your facility requires.
   j. Read the indicated voltage from the digital display of your meter.
      1. If the indicated voltage is correct for the amount of voltage that should be present in the circuit,

your meter is functioning normally and you may proceed with your test. Disconnect your test probes from the test circuit and skip the next step.

2. If the indicated voltage is not correct for the amount of voltage that should be present in the circuit, your meter is faulty and should not be used until repaired. Disconnect your test probes from the test circuit, and do not proceed until a properly working voltmeter can be obtained.

k. Connect the red and black test probes to the circuit you are troubleshooting. Remember that when measuring voltage, your test leads will be connected across, or in parallel with, the portion of the circuit you are testing.

l. Read the indicated voltage from the digital display of your meter. Your meter should indicate approximately zero volts if the circuit is indeed de-energized.

m. Carefully disconnect your meter from the circuit you are testing.

n. Reconnect your meter to the circuit with the similar known amount of voltage present that you previously used. This will verify that your meter is still functioning properly.

Review Questions

1. What are the two main causes of workplace accidents?

2. Explain why it can be dangerous to do a job for which you have not been trained.

3. What kind of safety information might be found on the label of a container of adhesive?

4. Name three safety devices that are found on construction tools or equipment.

5. How could a misunderstood message be a safety hazard?

6. How can prescription medicine be a safety hazard?

7. Name a working condition that might be a safety concern, and explain what you would do about it.

8. What is the overall purpose of OSHA?

9. What is MSDS an abbreviation for?

10. In your own words, state the five rules given in this chapter for handling hazardous materials.

11. What should you do if you find hazardous conditions on your work site?

12. List six items of PPE.

13. How much noise does it take to begin to damage your hearing?

14. What are the three parts of the fire triangle?

15. What are the four classes of fires, and what kinds of fire does each represent?

16. Name two classes of fires that should never be extinguished with water and explain why.

17. How can you tell which classes of fire a particular fire extinguisher can be used for?

18. Explain why working in a trench can be dangerous.

19. At what depth can the weight of soil be so much that it prevents a person from breathing?

20. What are two precautions that can make trench work safer?

21. What is the force called that causes electric current to move through a conductor?

22. Which has higher electrical resistance, wet wood or dry wood?

23. Describe the flow of electricity for a complete circuit when a person standing on wet ground touches an electric wire with faulty insulation.

24. List seven safety rules for electricity given in this chapter. You may refer to the chapter as you list these, but do not copy word for word from the chapter. Use your own words.

Activities

ACCIDENT REPORT

1. Search newspapers to find a report of a recent construction accident. If you do not subscribe to a newspaper or your paper does not describe a construction accident, you will find copies of many newspapers in the library and on the Internet. Make a copy of the article for your report.
2. Write a very brief description of your own, giving just the most important facts about the accident, including what caused it.
3. Explain what safety rules were broken to cause the accident and how it could have been prevented.
4. Give a brief report on the accident to your class.

SAFETY HAZARDS

Visit a construction site or base this activity on conditions in your school lab. Describe each of the safety hazards you see, explain why it is a hazard, and describe what has been done or should be done to correct the hazard. Find at least five hazards that either exist or have been corrected or avoided. Make a form like the one below to record your findings.

MATERIAL SAFETY DATA SHEET (MSDS)

Refer to the MSDS in Figure 3–10 to answer the following questions:

1. What is the product? _____
2. What phone number would you call if there was an emergency involving large amounts of this product? _____

   _____
3. At what temperature will this product ignite? _____ Would you say it is highly flammable or not? _____
4. What should be done if someone inhales the vapors of this product? _____

   _____
5. What PPE is recommended for those working with this material? _____

FIRE

Sketch a floor plan of your school shop or your job site, indicating where fire extinguishers are located. List the classes of fires that can be extinguished with each fire extinguisher on your sketch.

HAZARD	WHY A HAZARD	RECOMMENDATION TO CORRECT

MEASURING BODY RESISTANCE

Electric shock is a result of an electric current flowing through the body. The amount of current that flows through the body is determined by the makeup of the body (lean body tissue or fat), the entry and exit points on the body, the degree of moisture present on the skin, and the amount of voltage applied to the body. Using the ohmmeter function of a digital multimeter, you will measure your body resistance between several points. Then you will perform some simple calculations to determine the amount of voltage required to cause a fatal current to flow through your body.

DO NOT ATTEMPT TO PROVE THIS!

Materials Required:

• A digital multimeter with test leads

After completing this activity, you should understand the operation of a ground fault circuit interrupter.

PROCEDURE

1. Set your digital multimeter to the resistance range. (If you do not have an auto ranging digital multimeter, you will need to find the range that will yield the best readings.)
2. Insert the black test lead into the common jack on the digital multimeter.
3. Insert the red test lead into the volt/ohm jack on the digital multimeter.
4. Hold one test probe lightly in your right hand and the other test probe lightly in your left hand. Record your resistance measurement:
   From right hand to left hand _____ ohms.
5. Gradually increase the pressure with which you are holding the test probes. Record your resistance measurement:
   Increased pressure from right hand to left hand _____ ohms.
6. Hold one test probe in one hand and place the other test probe on your foot. Record your resistance measurement:
   From hand to foot _____ ohms.
7. Using a small amount of water, moisten your thumb and index finger of both hands. Now place one test probe between the thumb and index finger of your right hand. Place the other test probe between the thumb and index finger of your left hand. Record your resistance measurement:
   Moistened right hand to left hand _____ ohms.
8. Step 7 completed the measurements, and so turn your meter off.
   Perform the following calculations. Since we know that 0.1 ampere of current is considered to be fatal, we will calculate the amount of voltage required to cause 0.1 ampere of current to flow through various parts of your body. We will also see the effects when moisture is added. Again,

DO NOT ATTEMPT TO PROVE THIS!

We will use the formula volts = ohms $\times$ 0.1 ampere

1. Voltage from hand to hand (dry) from step 4:
   _____ volts
2. Voltage from hand to hand (squeezed tightly) from step 5:
   _____ volts
3. Voltage from hand to foot from step 6:
   _____ volts
4. Voltage from hand to hand (moist) from step 7:
   _____ volts

Analysis:

1. What happened to the resistance reading from hand to hand when you increased the pressure with which you were holding the test probes?
2. How does your resistance reading compare between your hand-to-hand measurement and your hand-to-foot measurement?
3. If you were to come in contact with an electric current, would you have a better chance of surviving if the current was allowed to flow from hand to hand, or from hand to foot? Explain your reasoning.
4. How does the resistance reading of your hand-to-hand measurement when your skin was dry compare with your hand-to-hand measurement when your skin was moist?
5. If you were to come in contact with an electric current, would you have a better chance of surviving if your skin was dry or moist? Explain your reasoning.

Chapter 4

Safety with Ladders, Scaffolds, and Machines

OBJECTIVES

After completing this chapter, the student should be able to:

- list safety considerations in erecting and working on scaffolds.
- choose the right ladder for a job and use it safely.
- list the safety considerations in working with internal combustion engines and engine-driven machines.
- list the safety considerations in working with compressed air and pneumatic tools.

Glossary of Terms

4:1 rule for ladders the rule that stipulates that ladders should be 1 foot away from the vertical surface against which they are placed for every 4 feet in height.

competent person one who is capable of identifying existing and predictable workplace hazards that are unsanitary, hazardous, or dangerous to employees and who has the authorization to take corrective measures to eliminate them.

portable ladder a ladder that can easily be picked up and carried to another location.

pump jack a device that attaches to a vertical pole and can be pumped up and down the pole. Pump jacks are used to support planks on which workers stand.

scaffolds temporary work platforms.

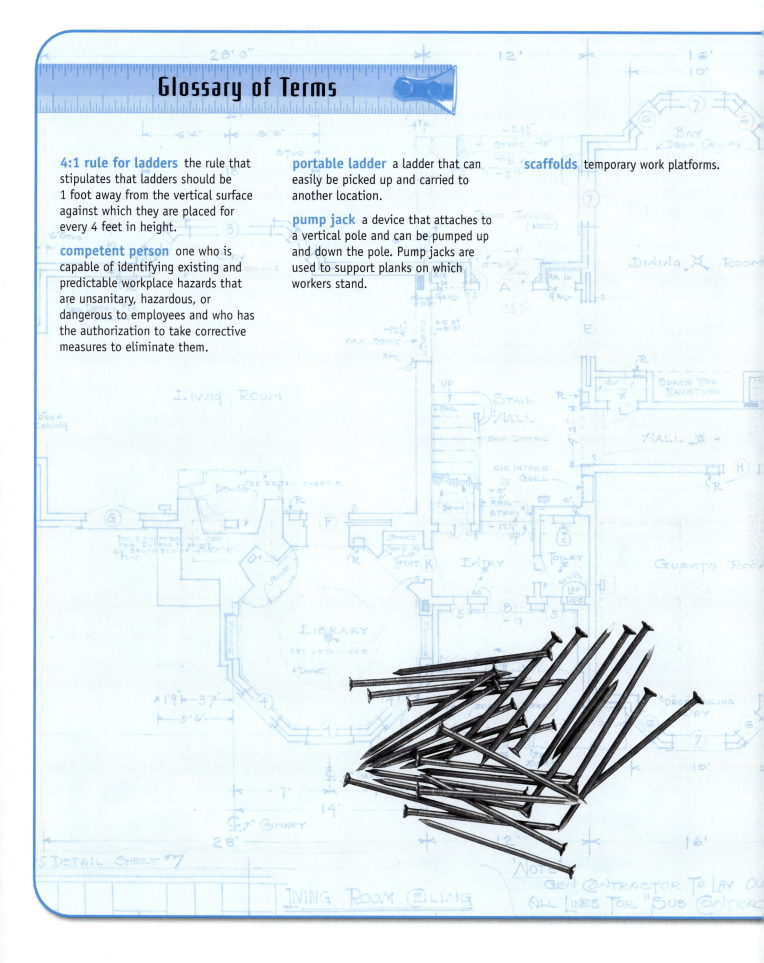

The topics discussed in this chapter are some of the biggest causes of serious injuries and even deaths on construction sites. Ladders and scaffolds present the risk of falls. Falling from a ladder or scaffold is bad enough, but sometimes heavy materials come down with the fall. Imagine the results when worker, ladder, and 50 pounds of copper tubing land in a heap. Electric and engine-driven machines present different risks, but they are just as serious.

Scaffolds

Scaffolds are temporary work platforms and are very widely used throughout the construction industry. Fabricated frame scaffolds (Fig. 4–1), are the most common type of scaffold because they are versatile, economical, and easy to use. They are frequently used in one or two tiers by residential contractors, painters, and so on, but their modular frames can also be stacked several stories high for use on large-scale construction jobs. The second most commonly used type of scaffold in residential construction is the **pump jack** (Fig. 4–2). Pump jacks are popular at sites where the work moves frequently. Pump jacks can be raised and lowered quickly by the pump mechanism. They can be moved from place to place quite quickly, because there are a small number of pieces.

According to OSHA, an estimated 2.3 million construction workers, or 65 percent of the construction industry, work on scaffolds frequently. Protecting these workers from scaffold-related accidents would prevent 4,500 injuries and 50 deaths every year, at a savings for American employers of $90 million in workdays not lost. In a recent Bureau of Labor Statistics study, 72 percent of workers injured in scaffold accidents attributed the accident either to the planking or support giving way or to the employee slipping or being struck by a falling object. These accidents could have been prevented by following OSHA requirements. The most important of those regulations is the requirement for all scaffolds to be built by someone who has been trained to build scaffolds by a "competent person." According to OSHA, "**competent person** means one who is capable of identifying existing and predictable hazards in the surroundings or working conditions which are unsanitary, hazardous or dangerous to employees and who has the authorization to take corrective measures to eliminate them." That training includes very specific topics, and a person who has been certified as a competent person knows how to erect scaffolds so that they will be safe work platforms. Never build your own scaffold if you have not been trained by a competent person, and never use scaffolds that were not built by a trained scaffold builder.

If you will be working on a scaffold, you should know some of the key elements of a safe scaffold presented in the list below. Even so, please keep in mind that knowing the key elements does not qualify you to build or inspect scaffolds.

- It is impossible for a stable structure to be built upon a foundation that does not start out square and level.

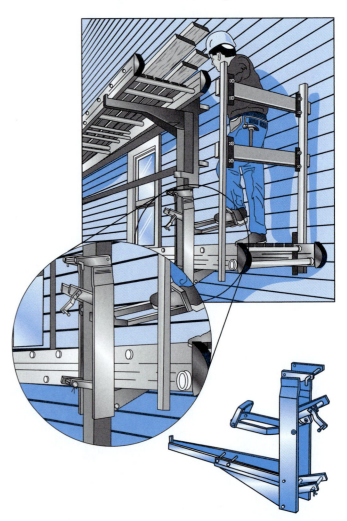

Figure 4–2 **Pump jack.** *Courtesy of Lynn Ladder & Scaffolding Company Inc.*

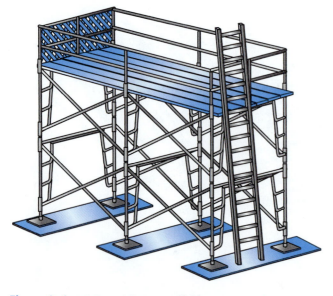

Figure 4–1 **Fabricated frame scaffold.**

OSHA has standards that apply specifically to the steps that must be taken to assure a stable scaffold base. The scaffold should be built on a firm base, with appropriate base plates and sills.

- Scaffold poles, frames, uprights, and so on, must be plumb and braced to prevent swaying and displacement.
- Frames and panels must be connected by cross, horizontal, or diagonal braces, alone or in combination, which secure vertical members together laterally.
- All brace connections must be secured to prevent dislodging.
- Employees must be able to safely access any level of a scaffold that is 2 feet above or below an access point.
- OSHA standards specifically forbid climbing cross braces as a means of access.
- Portable, hook-on, and attachable ladders must be positioned so as not to tip the scaffold.
- The number one scaffold hazard is worker falls. Protection consists of either personal fall-arrest systems or guardrail systems. These systems must be provided on any scaffold 10 feet or more above a lower level. Guardrail systems must be installed along all open sides and the ends of platforms.

Color-coded tags may be used to indicate when a scaffold was last inspected and what its condition was at the time (Fig. 4–3). A common system of scaffold tagging uses a green tag to indicate that the scaffold has been inspected and is safe for use. A yellow tag indicates that the scaffold may be used, but it does not meet all safety requirements and can only be used with a fall-protection harness. Finally, a red tag indicates that the scaffold is not safe for use. If tags are used, they are usually near the point where workers would access the scaffold. OSHA does not require tagging, and many scaffolds are inspected daily but are not tagged.

Portable Ladders

A **portable ladder** is a ladder that can easily be picked up and carried to another location. Stepladders and extension ladders are the most common portable ladders. They are used by all building trades. If they are maintained and used properly they are quite safe.

Choose the Right Ladder for the Job

Trying to reach beyond your safe limit or using a ladder at an improper angle because it was the wrong size can be the cause of an accident. If there is any chance that you will be working near electricity, choose a fiberglass ladder. Metal ladders and damp or oil-soaked wood ladders conduct electricity. Figure 4–4 will help you choose the right-size ladder.

Figure 4–3 Scaffold tags. *Courtesy of Seton Identification Products.*

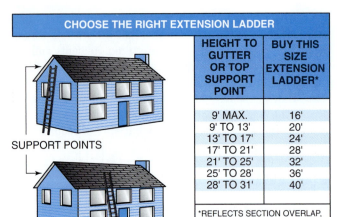

CHOOSE THE RIGHT EXTENSION LADDER		
	HEIGHT TO GUTTER OR TOP SUPPORT POINT	**BUY THIS SIZE EXTENSION LADDER***
SUPPORT POINTS	9' MAX.	16'
	9' TO 13'	20'
	13' TO 17'	24'
	17' TO 21'	28'
	21' TO 25'	32'
	25' TO 28'	36'
	28' TO 31'	40'
	*REFLECTS SECTION OVERLAP, LADDER ANGLE AND 3' EXTENSION ABOVE ROOF LINE.	

CHOOSE THE RIGHT STEPLADDER		
AVERAGE HEIGHTS	**MAXIMUM HEIGHT YOU WANT TO REACH**	**BUY THIS SIZE STEPLADDER**
	7'	3'
	8'	4'
	9'	5'
	10'	6'
	11'	7'
	12'	8'
	14'	10'
	16'	12'
	18'	14'
	20'	16'

Figure 4–4 **Choose the right ladder for the job.** *Courtesy of Werner Co.*

You should also consider the rating of the ladder. There are four industry ratings:

- Type IA Extra Heavy Duty 300 lb
- Type I Heavy Duty 250 lb
- Type II Medium Duty 225 lb
- Type III Light Duty 200 lb

The following safety rules should be obeyed when using any portable ladder:

Safe Climbing Habits

Courtesy of Werner Co.

"DOs"

- Read and carefully follow all instructions, warning labels, and manuals. Be aware of and comply with all federal, state, local, ANSI, OSHA and other codes and regulations.

- Keep your body centered on the ladder. Hold the ladder with one hand while working with the other hand whenever possible. Never let your belt buckle pass beyond either ladder rail.

- Move materials with extreme caution. Be careful pushing or pulling anything while on a ladder. You may lose your balance or tip the ladder.

- Get help with a ladder that is too heavy to handle alone. If possible, have another person hold the ladder when you are working on it.

- Climb facing the ladder. Center your body between the rails. Maintain a firm grip.
- Always move one step at a time, firmly setting one foot before moving the other.

- Haul materials up on a line rather than carry them up an extension ladder.
- Use extra caution when carrying anything on a ladder.

- Place an extension ladder at a 75½° angle. The set-back ("S") needs to be 1 ft. for each 4 ft. of length ("L") to the upper support point.

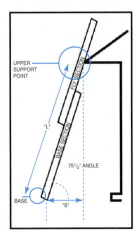

"DON'Ts"

- DON'T stand above the highest safe standing level.
- DON'T stand above the 2nd step from the top of a stepladder and the 4th rung from the top of an extension ladder. A person standing higher may lose their balance and fall.

- DON'T climb a closed stepladder. It may slip out from under you.
- DON'T climb on the back of a single-sided stepladder. It is not designed to carry a person's weight.

- DON'T stand or sit on a stepladder top or pail shelf. You could easily lose your balance or tip the ladder.
- DON'T climb a ladder if you are not physically and mentally up to the task.

- DON'T exceed the Duty Rating, which is the maximum load capacity of the ladder. Do not permit more than one person on a single-sided stepladder or on any extension ladder.

- DON'T place the base of an extension ladder <u>too close</u> to the building as it may tip over backward.
- DON'T place the base of an extension ladder <u>too far away</u> from the building, as it may slip out at the bottom. Set the ladder at a 75½° angle.

- DON'T overreach, lean to one side, or try to move a ladder while on it. You could lose your balance or tip the ladder. Climb down and then reposition the ladder closer to your work!

Internal Combustion Engines

Internal combustion engines are gasoline and diesel engines used to power generators, compressors, power trowels, and so on (Fig. 4–5). These engines can present several safety hazards:

- Fire hazard from fuel and exhaust
- Carbon monoxide poisoning from exhaust emission
- Powerful moving parts that can entangle clothing, hair, fingers, and hands
- Electric shock from the ignition system
- Hearing loss caused by loud engine operation

Following a few simple safety rules can prevent these injuries. These safety rules are easy to understand if you think about the hazards that may be present and the information provided earlier in this chapter:

- Do not operate engine-driven equipment until you have been instructed in its proper, safe use.
- Keep motor fuels in approved containers, made for that type of fuel.
- Keep open flames and sources of sparks away from fuel.
- Do not refuel a hot engine.
- Do not use gasoline or diesel fuel as a solvent or cleaning agent.
- Have a fire extinguisher for class B fires on hand and know how to use it.
- Never operate an internal combustion engine indoors.
- Keep engine-driven equipment in good condition. Do not remove guards, shields, or other devices intended to protect the user from moving parts.
- Keep engines dry.

Figure 4–5 Gasoline-powered generators are common on construction sites. *Courtesy of Guardian Ultra Source Portable Generator.*

- Do not operate an internal combustion engine without a working muffler, and if you must work near the engine, wear hearing protection.
- Remove the spark plug wire(s) before doing any repairs to the engine or attached machinery.

Compressed Air

Compressed air is used to power some tools such as pneumatic nailers. The air pressure used for these pneumatic tools is typically about 100 pounds per square inch but it might be as high as 200 pounds per square inch. At this pressure, compressed air can be dangerous. It can propel debris fast enough to puncture your skin, and it can do serious damage if directed at your eyes or ears. Wear safety glasses when using pneumatic tools. Compressed air must not be used under any circumstances to clean dirt and dust from clothing or off a person's skin. Most pneumatic tools used in construction have quick-disconnect fittings (Fig. 4–6). If a tool does not have this type of fitting, the air supply should be shut off before the tool is disconnected. Never attempt to use a pneumatic tool unless you have been trained in its proper use.

Caution must also be used when working around any other kinds of compressed gas. Compressed gas is found in oxygen and fuel cylinders for welding, brazing, and soldering. The

Figure 4–6 Quick-disconnect fitting for air hose.

oxygen for oxyacetylene welding is shipped at 2,500 pounds per square inch. Other welding fuels are also shipped at high pressures. Also, the refrigerant gas at some points in an air-conditioning system is under high pressure. Do not open a refrigeration system unless you have been properly trained and understand the gases and pressures in the system.

Review Questions

Select the most appropriate answer where applicable.

1. Who may build scaffolds?

2. What is the definition of a competent person?

3. Briefly list the eight safety rules for scaffolds given in this chapter. You may refer to the chapter as you list these, but do not copy word for word from the chapter. Use your own words.

4. Which ladder material should be used when you are working near electric lines?

    a. wood
    b. aluminum
    c. fiberglass
    d. wood and fiberglass are both okay

5. Describe the "belt buckle" rule for ladder use.

6. When is it acceptable to face away from a ladder while climbing it?

7. What should you do if you need both hands to carry something while climbing an extension ladder?

8. What is the highest step you should use on a step ladder?

9. If an extension ladder rests against a roof edge 20 feet from the ground, how far should the bottom be set back from the top?

10. Briefly list the 11 safety rules for internal combustion engines given in this chapter. You may refer to the chapter as you list these, but do not copy word for word from the chapter. Use your own words.

11. Which of the following is not a true statement?

    a. Compressed air can cause injury to the eyes.
    b. Compressed air can puncture human skin.
    c. Compressed air can cause injury to the ears.
    d. All of the above are true statements.

Activities

BUILDING A SCAFFOLD

Beginning workers in construction are often called upon to help build scaffolding. You should not attempt to build scaffolds on your own until you have been thoroughly trained by a competent person. In this activity you will work in a small group to build a safe scaffold. Do not climb on the scaffold you build until it has been inspected by your instructor.

Equipment and Materials

- All parts necessary to build fabricated metal scaffold two stages high
- 2-foot level
- Shovel
- 2 × 10 or 2 × 8 lumber as needed to prepare a level base
- Scaffold-grade wood planks or prefabricated scaffold planks

PROCEDURE

1. Working in the area designated by your instructor, lay out all scaffold parts and inspect them to ensure that they are in good condition.
2. If you are working on bare ground, you may need to prepare a level base. Shovel away high spots or build up low spots with 2 × 8 or 2 × 10 cribbing. Use a straight piece of lumber and a level to ensure that all four corners are level and solid (Fig. 4–7). If the cribbing does not make solid contact with the ground or if one corner is high or low, fix it before you go on.
3. Place adjustable base plates on two end frames, position them on the prepared base, and join the two end frames with two diagonal cross braces. It might be necessary to move or adjust the base you prepared so that the base plates fit squarely on top of the prepared surface. This step is especially important. If you start building your scaffold on a solid, level base, the rest is much easier. If your base is not level, the problem will be magnified as the scaffold goes up.
4. Check the end frames to see that the posts are plumb (use a level to ensure that they are

Figure 4–7 **Use lumber to prepare a level base for the scaffold bases.** *Courtesy of Lynn Ladder & Scaffolding Company Inc.*

perpendicular to the earth's surface—see page 000) and the horizontal pieces are level.

5. Have your instructor check your work before you proceed.
6. Plank the first stage of scaffolding. The planks should have cleats to prevent them from slipping off the end frame (Fig. 4–8).
7. Insert coupling pins in all four top corners, and set two more end frames on top of the bottom stage.
8. Install diagonal bracing on the second stage.
9. Place planking approximately 8 feet from the ground. The planking from the first stage can be moved up, or you may choose to leave that planking in place and place new planking on the second stage.
10. Install guardrails around your work platform, according to the manufacturer's instructions.

Note: Guardrails are not required for scaffolds below 10 feet, but they should be installed for this activity so that you can have that practice. Scaffolds of more than 10 feet require several more safety precautions such as securing the scaffold to the wall

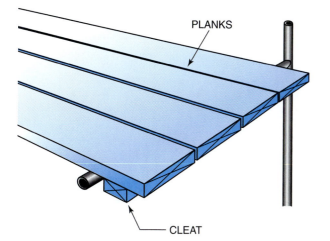

PLANKS

CLEAT

Figure 4–8 **The planking must be cleated or otherwise prepared so that it cannot slip off the scaffold frame.**

and implementing additional requirements for access to the scaffold.

11. Recheck to see that vertical poles are plumb and that horizontal supports are level.

12. Have your instructor inspect your scaffold before you climb or stand on it.
13. Make a list of all the safety precautions you followed in building your scaffold.

INSPECT A LADDER

A surprising number of ladders that are in use have serious safety hazards. You may have such a ladder at home, or your school might have one or more older ladders that present safety hazards.

PROCEDURE

Find a stepladder or extension ladder that has been in use for a few years and inspect it. Complete the following form as you do your inspection.

LADDER INSPECTION REPORT

Type of ladder (check one):

☐ stepladder
☐ straight, one-piece ladder
☐ extension ladder
☐ other

Material (check one):

☐ wood
☐ fiberglass
☐ aluminum

Are the manufacturer's safety labels in place and legible?

☐ yes
☐ no

What is the load rating of the ladder?

Are there any cracks, breaks, or repairs in the side rails?

☐ no
☐ yes (description)

Are there any cracks, breaks or repairs in the rungs or steps?

☐ no
☐ yes (description)

Is all hardware tight and in good condition?

☐ yes
☐ no (description)

Are feet in place and in good condition?

☐ yes
☐ no (description)

Is the ladder in good condition and safe to use within its load rating? Write a short statement about the ladder's overall condition.

Construction Math

$$P = 2l + 2w$$

SECTION THREE

CONSTRUCTION MATH

Success Stories

Jeff Haines

TITLE

Owner/President of Haines & Haines Incorporated, Syracuse, New York.

EDUCATION

Jeff graduated from State University of New York, College of Technology at Delhi, with an associate's degree in applied technologies.

HISTORY

By the 11th grade, Jeff was sure that he wanted to go into carpentry or cabinetmaking. Jeff really enjoyed technical classes, particularly in wood shop and technology. Even before he graduated from high school, Jeff bought substantial tools used in carpentry. Although many people learn a trade on the job, Jeff felt that a college education would provide more diversity and make him more marketable. His education helped him learn much more quickly once he was on the job. Jeff recalls, "In my early twenties I was teaching a coworker with more than 15 years of experience how to lay out stairs. This ability was a result of my education."

Before going into business for himself, Jeff worked for two reputable builders in the area. One specialized in building high-end homes and remodeling historic homes, where the emphasis was on craftsmanship and quality. His second employer was a development home-builder. Although they built quality homes, the emphasis was much more on production and profit margin. "These experiences (with two different types of builders) exposed me to two very different segments of the home-building/remodeling trades," states Jeff.

ON THE JOB

Jeff's company is small, with just two full-time employees and occasional part-time help. This means that much of the time Jeff is still in the field swinging a hammer. Jeff also meets with customers, prepares estimates, coordinates subcontractors, and orders materials. He enjoys the variety of responsibilities. "Every day is different. Every job is different. We could remodel 10 kitchens in a row, but no 2 would be alike. This variety stems not only from different styles of homes but also from the different people we work for."

In remodeling, it is important to talk with the homeowners enough to understand what they want the end result to be. "When we get done and they say, 'That's

just what I was looking for,' I have done my job." Jeff is thankful that he has skills that will always be needed. He encourages anyone who has an interest to give any of the skilled trades a try.

BEST ASPECTS

Jeff enjoys self-employment because it provides some flexibility in his work schedule. "It's nice to not have to ask the *boss* for time off if I need it," he says. The tradeoff, however, is that he often works an average of 50 to 60 hours a week. "All in all, I enjoy what I do," says Jeff, "and most of the time I don't mind the longer hours, I accept it as part of the responsibility of being a business owner."

CHALLENGES

Jeff is challenged by trying to find good quality help. He has two employees who are both dependable and talented in their trade. At one time he had four full-time employees, but the quality of his company's work slipped and profits dropped. In the future, if Jeff hires anyone, it will be on a trial basis so that he can ensure that the company maintains the level of quality that customers are used to and expect.

IMPORTANCE OF EDUCATION

"I don't know how to stress this enough. Even if you think all you want to do is saw 2 × 4s and hammer nails, GET AN EDUCATION. There is so much more to a skilled trade than I realized even after going through college. I still learn everyday," Jeff says. "It speaks volumes when you tell a prospective customer that you went to school for your trade. It tells people that you are serious about what you do." Jeff also suggests that if you decide to go to college for carpentry, and an electrical class is an elective requirement, do not take that class lightly. If the electrical subcontractor decides not to show up, your education could make you the most valuable employee on the job that day. "Learn all you can about all of the different trades you can. It will help you in the future."

FUTURE OPPORTUNITIES

Jeff's goal is to build an excellent reputation in custom kitchens. What sets his company apart is that it has the talent not only to do on-site remodeling, but also to construct custom cabinets. "Most companies do one or the other, but seldom both," says Jeff.

WORDS OF ADVICE

"Helping customers feel good about their homes is what this trade is all about. If you are only interested in running power tools and hammering nails, it will show in your work. If you make the trade about people, this will also show in your work."

Chapter 5 | Whole Numbers

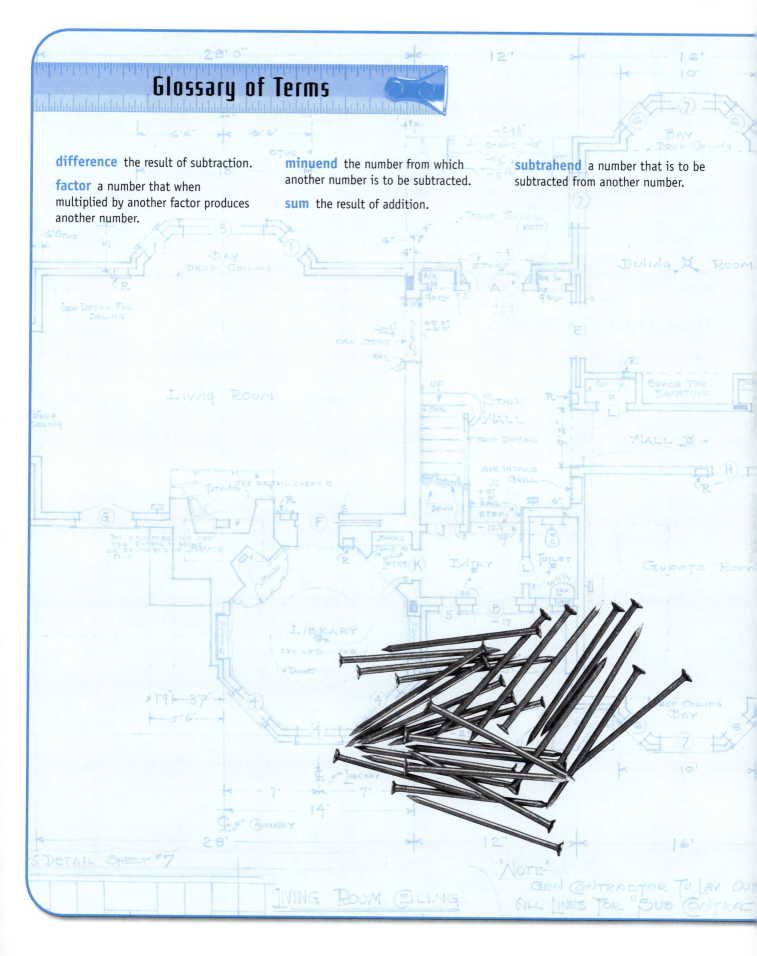

Glossary of Terms

difference the result of subtraction.

factor a number that when multiplied by another factor produces another number.

minuend the number from which another number is to be subtracted.

sum the result of addition.

subtrahend a number that is to be subtracted from another number.

hole numbers are used extensively throughout the construction industry and by people working in all the trades. It is necessary to be able to add, subtract, multiply, and divide whole numbers in order to figure amounts of materials, sizes of building components, and time and pay.

Basic Principles

The key to performing operations with whole numbers is memorizing number facts—addition tables, multiplication tables, and so on. It is also important to be sure that you line up columns of numbers correctly. The numerals on the right should line up over one another for addition, subtraction, and multiplication. If you are working with units, like feet, pounds, or quarts, be sure to label your answer with the units.

Addition Problems

Add the following quantities:

1. 13
   +8

2. 2,768
     814
     644
   +555

3. 270 lb.
    814 lb.
     58 lb.
   +9 lb.

4. 112 ft.
     96 ft.
     40 ft.
     57 ft.

5. 14 cu. yd.
    12 cu. yd.
   +5 cu. yd.

Solve the following problems:

6. How many feet of cove molding are needed to trim a room with walls of 16 feet, 12 feet, 16 feet, and 12 feet?

7. What is the total length of the house shown in the partial floor plan in Figure 5–1?

8. In a 5-day workweek, you worked 7 hours, 8 hours, 7 hours, 6 hours, and 9 hours. How many hours did you work that week?

9. A house requires ductwork in the following lengths: 26 feet, 18 feet, 9 feet, 8 feet, 8 feet, 5 feet, and 3 feet. How many total feet of ductwork are needed?

10. The branch circuits for a house are listed in Figure 5–2. What is the total current load for the house?

Bedroom lighting	15 amps
Living room, den, hall, and kitchen lighting	15 amps
Bedroom outlets	15 amps
Living room, den, hall outlets	15 amps
Kitchen outlets, circuit 1	20 amps
Kitchen outlets, circuit 2	20 amps
Bathroom 1	20 amps
Bathroom 2	20 amps
Kitchen range	50 amps
Total	

Figure 5–2

11. A house requires 12 cubic yards of concrete for the main slab, 6 cubic yards for the footings, and 7 cubic yards for the garage. How many total cubic yards are required for the house?

12. A house uses 480 feet of 12 AWG cable, 210 feet of 14 AWG, 31 feet of 10 AWG, and 15 feet of 2/0 AWG. What is the total amount of cable used?

13. A heating unit has to provide the following amounts of heat to four rooms: 28,000 BTUs, 21,000 BTUs, 12,000 BTUs, and 17,000 BTUs. What is the total amount of heat the unit must produce?

14. A house needs 6 gallons of flat white paint, 8 gallons of flat ivory paint, 2 gallons of semigloss white paint, and 1 gallon of light tan semigloss paint. How many gallons of paint are needed?

15. What is the total distance around a rectangular concrete pad that is 28 feet wide and 42 feet long?

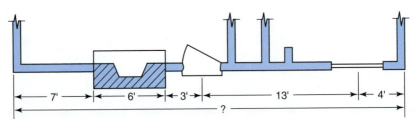

Figure 5–1

Subtraction Problems

Subtract the following quantities:

16. 29
    −5

17. 2,768
    −814

18.   57 ft.
    −18 ft.

19.   114 lb.
    −102 lb.

20.   28 cu. yd.
    −13 cu. yd.

Solve the following problems:

21. If you cut a piece 106 inches long from a 120-inch board, how many inches are left?

22. From a supply of 2,643 feet of tubing, you use 114 feet. How much tubing is left in the supply?

23. If 145 cubic yards of earth need to be excavated for a house foundation and 70 cubic yards can be used as fill on the site, how much must be disposed of elsewhere?

24. Figure 5–3 shows a floor plan for a detached garage. What is the width of wall space A at the front of the garage?

25. In Figure 5–3, what is the distance from the outside of the rear wall of the garage to the center of the side door opening, dimension B?

26. What is the length of missing dimension C at the rear of the garage in Figure 5–3?

27. Working from a 500-foot coil of electrical cable, an electrician uses 46 feet, 41 feet, 28 feet, 14 feet, 27 feet, and 19 feet. How much cable remains on the coil?

28. Find missing dimension A in Figure 5–4.

29. Find missing dimension X in Figure 5–5.

30. A sling rated for 1,500 pounds is used to hoist a 536-pound air conditioner into place. What is the margin of safety for the sling?

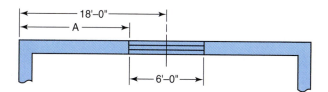

Figure 5–4

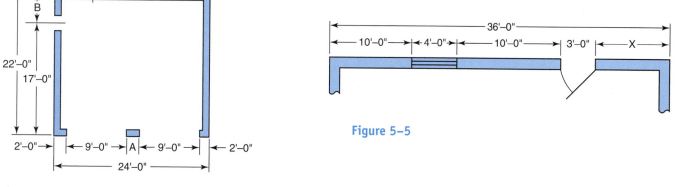

Figure 5–5

Figure 5–3

Multiplication Problems

Multiply the following quantities:

31.   9
    ×5

32.   22
    ×3

33.   427 ft.
    ×23

34.   377 gal.
    ×14

35.   54°
    ×4

Solve the following problems:

36. What is the total length of 24 pieces of door casing if each piece is 7 feet long?

37. If you work 37 hours at $12 per hour, what should your pay be?

38. If you need to make electrical connections in 27 boxes and each box requires 3 wire nuts, how many wire nuts do you need?

39. If you need to charge 8 air conditioners and each one takes 6 pounds of refrigerant, how many total pounds of refrigerant will you need?

40. How many feet of cable are there in 5 coils if each coil contains 250 feet?

41. If 9 lightbulbs consume 75 watts each, how many total watts are consumed?

42. If you need 8 pieces of duct and each piece needs to be 4 feet long, how many feet of duct do you need?

43. Every 10 feet of pipe requires 3 strap hangers, and every strap hanger requires 2 screws. How many screws are needed for 30 feet of pipe?

44. How many cubic yards of earth are moved if a dump truck carrying 9 yards makes 7 trips?

45. If 15-amp circuit breakers cost $9 a piece, what is the total cost of 14 circuit breakers?

Division Problems

Divide the following quantities:

46. $4\overline{)12}$

47. $6\overline{)144}$

48. $8\overline{)96 \text{ ft.}}$

49. $14\overline{)5,012 \text{ lb.}}$

50. $120\overline{)14,400 \text{ V}}$

Solve the following problems:

51. How many fixtures requiring 4 fluorescent lamps each can be supplied from a case of 24 fluorescent lamps?

52. How many 4-foot pieces can be cut from a 12-foot 2 × 4?

53. If one person can apply 300 square feet of insulation per hour, how long will it take that person to apply 3,300 square feet of insulation?

54. If a pipe requires a hanger every 4 feet, how many hangers will be needed for 28 feet of pipe? (Hint: No hanger is needed at either extreme end. See Fig. 5–6.)

55. A room with 72 lineal feet of wall space requires an electric outlet every 12 feet. How many outlets are required for the room?

56. A cylinder contains 352 ounces of refrigerant. How many pounds does it contain? (There are 16 ounces in a pound.)

57. For a gas furnace to maintain an air-to-gas ratio of 12 to 1 (it burns 12 cubic feet of air for every 1 cubic foot of gas), how much gas is required to mix with 600 cubic feet of air?

58. A lighting circuit can supply 1,200 watts of incandescent light. How many 150-watt fixtures can it support?

59. If you work 40 hours and are paid $720 for your work, what is your hourly pay?

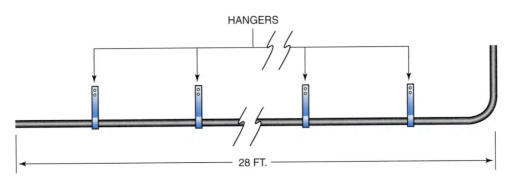

Figure 5–6

Combined Operations

In this section you will practice solving problems that combine addition, subtraction, multiplication, and division of whole numbers. In solving these problems, start by reading the problem carefully to see what information is given and what must be determined. Identify the operations that must be performed and the order in which they must be performed.

Problems

Solve the following problems:

60. A carpenter cuts 4 pieces of weather stripping from a roll 180 inches long. How much remains on the roll if the pieces were 40 inches, 40 inches, 34 inches, and 34 inches?

61. A contractor hired 3 plumbers to work 9 hours per day for 5 days and 2 laborers to work 7 hours for 1 day. What was the total number of hours worked?

62. How many bundles of shingles are needed to cover a two-part roof if one part is 400 square feet and the other part is 600 square feet? It takes 3 bundles of shingles to cover 100 square feet.

63. The electrical code requires that no point on a wall can be more than 6 feet from an electric outlet. How many electric outlets are required for the rooms shown in Figure 5–7? Do not allow for wall thickness.

64. A cylinder of refrigerant weighs 24 pounds 4 ounces. If you use 28 ounces to charge a system, what is the weight of the cylinder when you finish? (Hint: Change the pounds and ounces to ounces. There are 16 ounces in a pound.)

65. If a coil of nails will apply 5 sheets of plywood and each sheet of plywood is 32 square feet, how many coils of nails will be needed to apply all the plywood on a 16-foot by 40-foot roof? (Hint: Multiply the width of the roof by its length to find the area in square feet.)

66. In 1 month a plumbing contractor uses 1,200 feet of 1/2-inch copper pipe at $8 per 20-foot length, 450 feet of 3/4-inch copper pipe at $11 per 20-foot length, 310 feet of 1-inch copper pipe at $9 per 10-foot length, 510 feet of 4-inch plastic pipe at $4 per 10-foot length, and 264 feet of 1½-inch plastic pipe at $3 per 10-foot length. What is the total amount he paid for pipe for that month?

67. A manufacturer recommends one lighting fixture for every 80 square feet of ceiling area. In a room with a suspended ceiling, 4 tiles wide by 15 tiles long, how many fixtures should be installed? (Each tile is 8 square feet.)

68. If you cut 13 pieces, each 40 inches long, from a supply of 5 pieces of lumber, each 120 inches long, how much lumber remains?

69. If 2 people can install 75 square feet of siding per hour, how long will it take 4 people working at the same speed to side a house with walls of 16 feet × 10 feet, 12 feet × 10 feet, 20 feet × 18 feet, and 13 feet × 20 feet?

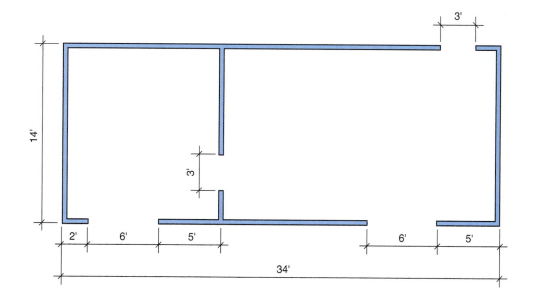

Figure 5–7

Chapter 6 Decimals

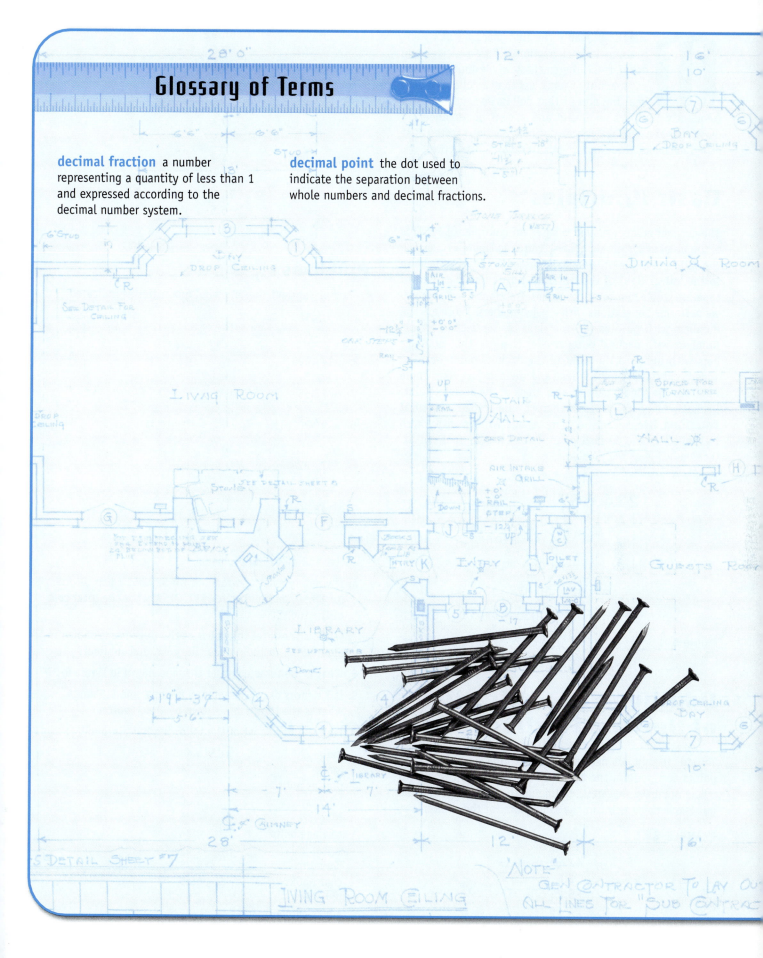

Glossary of Terms

decimal fraction a number representing a quantity of less than 1 and expressed according to the decimal number system.

decimal point the dot used to indicate the separation between whole numbers and decimal fractions.

oney is probably the first place all of us encounter decimal fractions. If we consider one dollar to be the smallest whole number, then any amount of cents less than a dollar is a decimal fraction of that dollar. Decimals are also frequently used by engineers and shown on engineering drawings. You will encounter decimal fractions, often just called decimals, throughout your study of construction.

Basic Principles

The decimal number system is based on multiples of 10. This is the number system we generally use. As Figure 6–1 shows, the values of digits (numerals) are based on how far they are to the left or right of the **decimal point** (the dot used to separate whole numbers and decimal fractions). In Chapter 5 we performed operations with whole numbers. Those are the numbers with no digits to the right of the decimal point. Numbers with decimal parts smaller than 1—that is, numbers with digits to the right of the decimal point—can be added, subtracted, multiplied, and divided just like whole numbers. We simply have to keep track of the decimal point. Decimal numbers of less than 1 are called **decimal fractions.** Decimal fractions are usually written with a zero to the left of the decimal point, but that zero does not affect the value of the number.

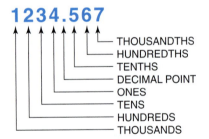

1234.567

- THOUSANDTHS
- HUNDREDTHS
- TENTHS
- DECIMAL POINT
- ONES
- TENS
- HUNDREDS
- THOUSANDS

Figure 6–1

Addition and Subtraction of Decimal Fractions

To add or subtract decimal numbers, line up the decimal points and perform the operation exactly as you would for whole numbers. Be careful to line up the decimal points and put the decimal point in the same place for the answer.

EXAMPLE ①

```
  41.12
   2.6
 + 5.25
-------
  48.97
```

```
  56.92
   7.8
+ 61.23
-------
 125.95
```

```
   9.4
 - 5.2
------
   4.2
```

```
  24.3
 - 5.6
------
  18.7
```

Problems

Add or subtract the following quantities as indicated:

1.
```
  13.52
 + 8.41
```

2.
```
 141.625
 372.2
 + 27.55
```

3.
```
  47.58
 - 5.31
```

4.
```
 573.228
 + 37.56
```

5.
```
 148.38
-139.78
```

Solve the following problems:

6. If you buy a set of wrenches for $29.95, a pair of pliers for $8.50, and a level for $22.49, what is the total cost of the tools?

7. What is the total thickness of the floor frame shown in Figure 6–2, including the sill, joist, and subfloor?

8. What is the total length of the pipe shown in Figure 6–2?

9. If 3 motors draw 2.2 amps, 5.8 amps, and 8.5 amps, what is the total amps drawn by the 3 motors?

10. Find the total cost of the following hardware bill:

nails	$16.40
locks	29.44
drawer pulls	7.40
latches	5.00
hinges	4.75
carriage bolts	4.34
screening	11.89
doorstops	3.75
coat hooks	6.60
handles	7.65
elbow catches	0.89

11. A control circuit has resistances of 253.7 ohms, 15.1 ohms, and 474.6 ohms in series. What is the total resistance of the circuit?

12. How many squares of shingles are needed to cover the roof shown in Figure 6–3?

13. If you recharge four air conditioners, using 6.7 pounds of refrigerant in one, 11.3 pounds in another, 13.7 pounds in the third, and 8.9 pounds in the fourth air conditioner, how much total refrigerant is used?

14. If you have 700 square feet of insulation and use 456.7 square feet, how much do you have left?

15. If you have 8.5 gallons of white paint and you estimate that a job will require 3.75 gallons, how much will you have left?

16. If you use 9.25 pounds of nails from a supply of 21.75 pounds, what is left in the supply?

17. A plumber receives $1,400 for installing a vanity and sink. If the materials cost $979.55, what is left for labor and profit?

18. What is the total cost of construction of the house shown in the following table?

Table for Questions 18 and 19

excavation and earthwork	$2,760.58
concrete and masonry	$4,233.50
framing	$29,450.78
electrical	$7,689.90
plumbing	$5,400.00
HVAC	$5,447.62
insulation	$1,345.97
drywall	$3,145.45
finish carpentry and trim	$7,785.91
painting	$2,970.25
landscaping	$850.00

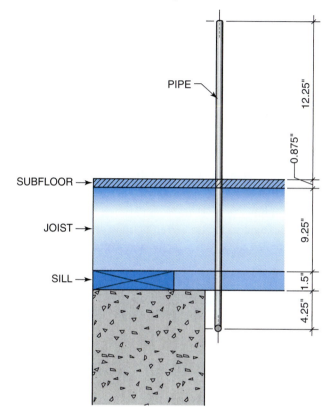

Figure 6–2

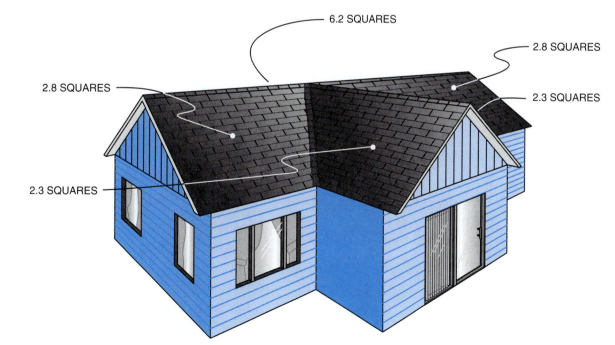

Figure 6–3

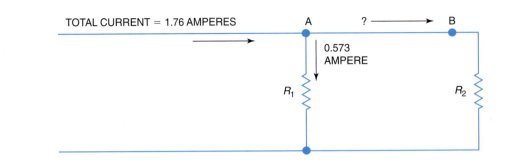

TOTAL CURRENT = 1.76 AMPERES

0.573 AMPERE

R_1 R_2

A   ?   B

Figure 6–4

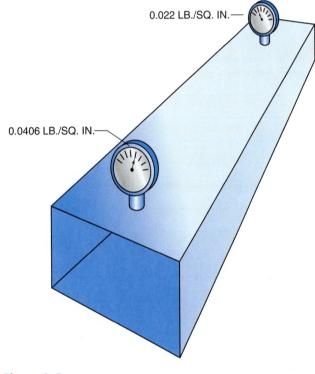

0.022 LB./SQ. IN.

0.0406 LB./SQ. IN.

Figure 6–5

19. What is the profit if the contractor receives $90,000 for the house in question 18?

20. You pay $2,915.45 for materials and $300 for labor to install an air conditioner. What is your profit if you are paid $3,850 for the job?

21. If you have an allowance of $1,500 for light fixtures and you buy light fixtures costing $121.20, $179.80, $54.95, $247.45, $35.95, and $35.95, how much of the allowance remains?

22. The field current of a motor is 1.32 amps at no load. When the full load is applied, the current increases to 1.93 amps. What is the difference in amps from no load to full load?

23. The total current in Figure 6–4 divides at point A. How much current exists at point B?

24. At 150°F, dry air has a density of 0.0652 pound per cubic foot. At 70°F, it has a density of 0.075 pound per cubic foot. What is the difference in density between air at 70°F and 150°F?

25. Determine the pressure drop between the ends of the duct in Figure 6–5 in pounds per square inch.

Multiplication of Decimal Fractions

To multiply decimals it is not necessary to line up the decimal points. Just line up the digits on the right, as you would for whole numbers. After you complete the multiplication, count the total number of digits to the right of the decimal point in both rows. (In the example below, there is 1 digit to the right in the top row and 3 digits to the right in the bottom row for a total of 4 digits to the right of the decimal point.) Place the decimal point in the answer that many places from the far right.

EXAMPLE ❷ Multiply 42.4 by 0.482.

```
       42.4
    × .482    (Zeros to the left of all other digits and the
       848    decimal point can be ignored.)
      3392
      1696
    20.4368   (Decimal point is 4 places from the far right.)
```

When multiplying some decimal fractions, the product (the answer) will have a smaller number of digits than the number of decimal places required. For these products, add as many zeros to the left of the product as necessary to give the required number of decimal places.

EXAMPLE ❸ Multiply 0.27 by 0.18.

```
  0.27   (2 places)
×0.18   (2 places)
  216
  27
0.0486   (One zero was added to the left of the 4, so
         there would be 4 places.)
```

When multiplying or dividing some decimal fractions, the answer can contain several more decimal places than the original problem contained. To show all these decimal places in the answer would suggest that the answer is more accurate than the original problem and given information. To avoid this, results of multiplication and division are usually rounded off to the same number of decimal places as the highest number of places used in the information given in the problem. Thus if the problem has 3 decimal places, the answer should be rounded off to 3 places.

To round a decimal fraction, locate the digit in the number that gives the desired number of decimal places. Increase that digit by 1 if the next digit to the right is 5 or more. Do not change the value if the digit to the right is less than 5. Drop all digits that follow.

EXAMPLE ❹ To round 0.63861 to 3 decimal places, increase the 8 to 9 (the next digit is 5 or more) and drop the 61. The rounded number is 0.639

To round 3.0746 to 2 places, do not change the 7 in the second place (the next digit is less than 5), but do drop the 46. The rounded number is 3.07

Problems

Multiply the following quantities:

26. 29.3
     × 5

27.   27.68
     × 81.4

28.   1.414
     ×0.25

29.   0.12
     × 0.22

30.   29.3
     × 0.115

Solve the following problems:

31. The rating stamped on an oil burner nozzle indicates how many gallons of oil are sprayed each hour. If the nozzle is marked 0.65 and it sprays for 8.75 hours, how many gallons are sprayed?

32. If 0.8 square of shingles can be applied in 1 hour, how long will it take to apply 8.25 squares?

33. What is the resistance of a piece of copper wire that has a size of 2.5 mil feet if 1 mil foot has a resistance of 10.4 ohms?

34. A certain size of lumber costs $1.15 per board foot. What is the cost of 50.5 board feet of that lumber?

35. What is the cost of 1,450 feet of BX cable if it sells for $0.675 per foot?

36. If a truck uses 0.9 gallon per mile and gasoline costs $1.519 per gallon, what does it cost to drive the truck 13,215 miles?

37. How much larger is the diameter of number 11 wire than the diameter of number 16 wire. Refer to Figure 6–6.

38. Sometimes in the manufacture of wire, the wire is actually slightly larger or smaller than intended. If the wire is measured and found to be 0.0799 inch in diameter, what standard wire size is it intended to be? Refer to Figure 6–6.

39. Pipe is made in many weights, the most common being schedule 40, schedule 80, and schedule 120. Figure 6–7 shows the inside diameters of 1-inch pipe. (Nominal means it is sold as 1-inch pipe, although its actual size may be different.) What is the difference in inside diameter of 1-inch schedule 40 pipe and 1-inch schedule 80 pipe?

40. According to Figure 6–7, what is the wall thickness of 1-inch schedule 80 pipe?

American Wire Gauge Size Number	Wire Diameter In Inches
10	0.10190
11	0.09074
12	0.08081
13	0.07196
14	0.06408
15	0.05707
16	0.05082

Figure 6–6

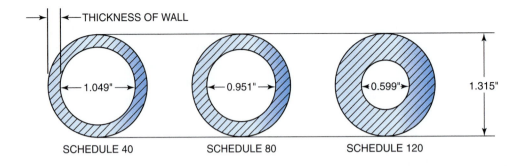

THICKNESS OF WALL

←1.049"→   SCHEDULE 40

←0.951"→   SCHEDULE 80

←0.599"→   SCHEDULE 120

1.315"

Figure 6–7

Division of Decimal Fractions

To divide decimal fractions, use the same procedure as with whole numbers. Move the decimal point of the divisor to the right as many places as necessary to make it a whole number. Move the decimal point of the dividend the same number of places to the right. Add zeros to the dividend as necessary to allow you to move the decimal point. Place the decimal point in the answer directly above the decimal point in the dividend. Divide as with whole numbers. Zeros may be added to the dividend to make the number of decimal places required for the answer.

EXAMPLE **5** Divide 0.6150 by 0.75.

Move the decimal point 2 places to the right in the divisor.

Move the decimal point 2 places in the dividend.

Place the decimal point in the answer directly above the decimal point in the dividend.

Divide as with whole numbers.

$$
\begin{array}{r}
0.82 \\
\text{Divisor} \rightarrow 0\,75.\,)\overline{0\,61.50} \leftarrow \text{Dividend} \\
60\ 0 \\
\hline
1\ 50 \\
1\ 50 \\
\end{array}
$$

Divide 10.7 ÷ 4.375. Round the answer to 3 decimal places.

Move the decimal point 3 places to the right in the divisor.

Move the decimal point 3 places in the dividend, adding 2 zeros.

Place the decimal point in the answer directly above the decimal point in the dividend.

Add 4 zeros to the dividend. One more zero is added than the number of decimal places required in the answer.

Divide as with whole numbers.

$$
\begin{array}{r}
2.4457 \approx 2.446 \\
4\,375.\,)\overline{10\,700.0000} \\
8\ 750 \\
\hline
1\ 950\ 0 \\
1\ 750\ 0 \\
\hline
200\ 00 \\
175\ 00 \\
\hline
25\ 00 \\
21\ 875 \\
\hline
3\ 1250 \\
3\ 0625 \\
\hline
625 \\
\end{array}
$$

Problems

Divide the following quantities:

41. 144
     ÷ 0.6

42. 96 ft.
     ÷1.5

43. 4.85 gal.
     ÷ 6

44. 10.278 in.
     ÷ 0.25

45. 1.447 yd.
     ÷6.11

Solve the following problems:

46. In Figure 6–8, each of 23 courses of siding is equally spaced. How much of each piece of siding is showing?

47. How many pieces of 2.25-inch flooring will be required to cover a closet floor that is 34.75 inches wide?

48. A certain size wire has a resistance of 3.65 ohms per 1,000 feet. What is the resistance of 10 feet of the wire?

49. If 500 hundred feet of type NM cable costs $49.65, what is the cost of 270 feet of this cable?

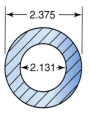

Figure 6–9

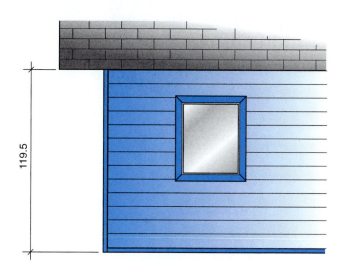

119.5

Figure 6–8

50. An oil burner used 6.68 gallons of oil in 4.25 hours. How many gallons did it use per hour?

51. What is the wall thickness of the pipe shown in Figure 6–9?

52. A box of 12 dimmer switches costs $38.64. What is the cost of a single switch?

53. The density of a substance is its weight divided by its volume. If 4.5 cubic feet of R-134a refrigerant weighs 378.23 pounds, what is its density in pounds per cubic foot?

54. If it takes 6.5 hours to install 54.75 feet of air-conditioning duct, how much time is used to install 1 foot?

55. The weight of a certain wire is 145 pounds per 1,000 feet. What is the cost of 600 feet of this wire at $1.40 per pound?

Chapter 7 | Fractions

After completing this chapter, the student should be able to:

- add fractions.
- subtract fractions.
- multiply fractions.
- divide fractions.

Glossary of Terms

denominator the number on the bottom of a common fraction.

fraction bar the horizontal line in a common fraction.

lowest common denominator the smallest denominator that can be used to express all of the common fractions in a set.

mixed number a number made up of a whole number plus a common fraction.

numerator the number on the top of a common fraction.

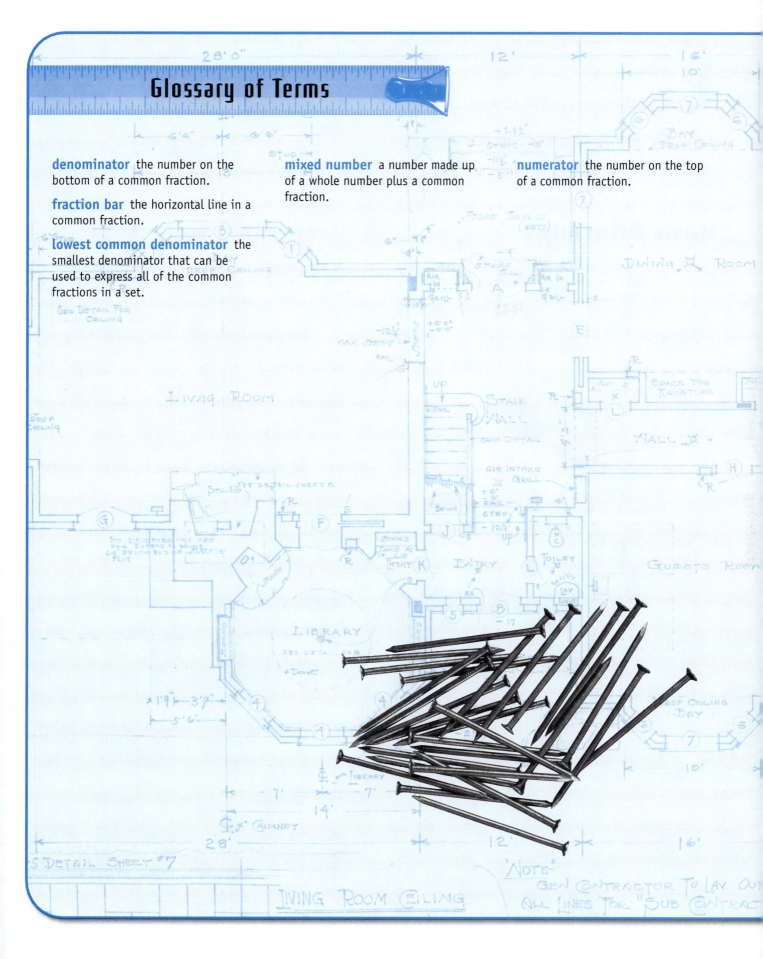

Fractions are commonplace in construction. In the Customary System of measurements we use most often in America, we use fractions of an inch to express measurements of distances. Common fractions, frequently just called fractions, are used for many other things, such as parts of volume measure (1/2 cup) or parts of a dollar (a quarter). You will need to be comfortable with adding, subtracting, multiplying, and dividing common fractions and converting common fractions to decimal fractions and converting decimal fractions to common fractions.

Basic Principles

A fraction is a part of a whole. A whole inch may be divided into equal parts in many ways. For instance, an inch may be divided into eight equal parts (Fig. 7–1). Each part is an eighth of an inch (1/8 inch). If five of these parts were needed in measuring a length, the quantity would be five-eighths of an inch and would be written as 5/8 inch.

Figure 7–1 Eighths of an inch.

In a fraction, the number of parts is called the **numerator** → $\dfrac{5}{8}$ ← The whole is called the **denominator.**

Sometimes it is necessary to change fractions to *equivalent fractions*. Equivalent fractions are fractions that have the same value. The value of a fraction is not changed when the numerator and denominator are both multiplied or divided by the same number.

EXAMPLE ① Express 5/8 as thirty-seconds. (5/8 = ?/32)

First determine what number the denominator of the original fraction is multiplied by to get the desired denominator. You can do this by dividing the desired denominator by the original denominator. (32 ÷ 8 = 4)

Next multiply the numerator and the denominator by your answer to the division problem, in this case 4. (5/8 × 4/4 = 20/32)

Often when working with two or more fractions, it is necessary to find the **lowest common denominator.** The lowest common denominator is the smallest denominator that is evenly divisible by each of the denominators of the fractions.

EXAMPLE ② The lowest common denominator of 3/4, 5/8, and 13/32 is 32, because 32 is the smallest number evenly divisible by 4, 8, and 32. (32 ÷ 4 = 8, 32 ÷ 8 = 4, and 32 ÷ 32 = 1)

EXAMPLE ③ The lowest common denominator of 2/3, 1/5, and 7/10 is 30, because 30 is the smallest number evenly divisible by 3, 5, and 10. (30 ÷ 3 = 10, 30 ÷ 5 = 6, and 30 ÷ 10 = 3)

If a common denominator is not readily apparent, one can be found by the following procedure:

1. Factor every denominator into its smallest factors. Factors are numbers used in multiplying. For example, 3 and 5 are factors of 15.

2. Find the product of all the different factors, using each factor the greatest number of times it is found in a denominator. This product is the least common denominator.

3. Multiply the original fraction—both the denominator and the numerator—by the factor of the least common denominator that was missing in the original denominator.

EXAMPLE ④ Express 5/6 and 4/9 as equivalent fractions with their lowest common denominator.

$$6 = 2 \times 3$$
$$9 = 3 \times 3$$

The different factors are 2 and 3. The factor 3 appears twice in 9, and so the least common denominator is 18, or $2 \times 3 \times 3$.

Multiply the original fractions as follows:

$$\frac{5}{6} = \frac{5}{2 \times 3} = \frac{5}{2 \times 3} \times \frac{3}{3} = \frac{15}{18}$$

$$\frac{4}{9} = \frac{4}{3 \times 3} = \frac{4}{3 \times 3} \times \frac{2}{2} = \frac{8}{18}$$

When working problems with fractions, the answer should be expressed in its lowest terms. A fraction is in its lowest terms when the numerator and denominator do not contain the same factor.

EXAMPLE ⑤ Express 12/16 in its lowest terms.

Determine the largest factor for both the numerator and the denominator. (Since 12 and 16 can both be divided by 4, you can divide the numerator and denominator by 4.)

$$\frac{12 \div 4}{16 \div 4} = \frac{3}{4}$$

Adding Fractions

To add fractions, express them with their least common denominators, add the numerators, and then write the sum of the numerators over the common denominator.

Problems

Add the following fractions:

1. 1/4 + 3/4
2. 3/8 + 1/4
3. 1/2 in. + 1/4 in. + 1/8 in.
4. 3/32 gal. + 1/5 gal. + 1/2 gal.
5. 1/8 hr. + 1/12 hr. + 1/3 hr.

Adding Mixed Numbers

A **mixed number** is a whole number plus a fraction, such as $6\frac{2}{3}$. In certain fractions the numerator is larger than the denominator. To express the fraction as a mixed number, divide the numerator by the denominator. Express the fractional part in lowest terms.

EXAMPLE 6 Express 38/16 as a mixed number.

1. Divide the numerator by the denominator. ($38 \div 16 = 2^{6}/_{16}$)
2. Express the fractional part in lowest terms. $\dfrac{6 \div 2}{16 \div 2}$ = 3/8
3. Combine the whole number and fraction. $2^3/_8$

To express a mixed number as a fraction, multiply the whole number by the denominator of the fractional part. Add the numerator of the fractional part. The sum is the numerator of the fraction. The denominator is the same as the denominator of the original fractional part.

EXAMPLE 7 Express $7^3/_4$ as a fraction.

1. Multiply the whole number by the denominator of the fractional part. $7 \times 4 = 28$
2. Add the numerator of the fractional part to the product found in step 1. $3 + 28 = 31$
   The sum (31) is the numerator of the fraction.
3. Make the denominator of the new fraction the same as the denominator of the original fraction. $7^3/_4 = 3^1/_4$

Problems

Add the following and express the answers in their lowest terms:

6. $2^7/_8 + 1^3/_4$
7. $3^3/_4$ gal. + $5^2/_3$ gal. + $6^5/_8$ gal.
8. 7/16 in. + 7/8 in. + 19/32 in.
9. 11/12 lb. + $1^5/_8$ lb.
10. 5/8 W + 9/16 W + $2^{13}/_{16}$ W

Subtracting Fractions

Fractions must have a common denominator to be subtracted. To subtract a fraction from a fraction, express the fractions as equivalent fractions having a common denominator. Subtract the numerators. Write their difference over the common denominator.

EXAMPLE 8 Subtract 3/4 from 15/16.

Express the fractions as equivalent fractions having a common denominator of 16.

$$15/16 = 15/16$$
$$3/4 = 12/16$$
$$15 - 12 = 3$$
$$15/16 - 12/16 = 3/16$$

To subtract a fraction or a mixed number from a whole number, express the whole number as an equivalent mixed number. The fraction of the mixed number should have the same denominator as the denominator of the fraction that is subtracted. Subtract the numerators of the fractions and write their difference over the common denominator. Subtract the whole numbers. Combine the whole number and the fraction and express the answer in lowest terms.

EXAMPLE 9 Subtract $5^{15}/_{32}$ from 12.

1. Express the whole number as an equivalent mixed number with the same denominator as the denominator of the fraction that is to be subtracted. $12 = 11^{32}/_{32}$
2. Subtract the fractions. $32/32 - 15/32 = 17/32$
3. Subtract the whole numbers. $11 - 5 = 6$
4. Combine the whole number and the fraction. $6^{17}/_{32}$

Problems

Subtract the following quantities and express the answers in lowest terms:

11. 3/4 − 1/4
12. 7/8 − 1/26
13. $1^2/_3$ ft. − 1/2 ft.
14. 14 tons − $2^5/_8$ tons
15. $95^1/_2$ in. − $14^{15}/_{32}$ in.

Solve the following problems involving addition and subtraction:

16. Six resistances are connected in series, as shown in Figure 7–2. What is the total resistance of the circuit?

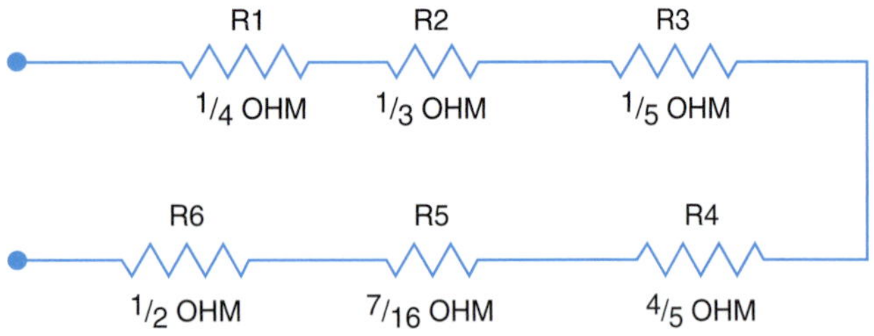
R1
R2
R3
1/4 OHM
1/3 OHM
1/5 OHM
R6
R5
R4
1/2 OHM
7/16 OHM
4/5 OHM

Figure 7–2

17. What would be the resistance of the circuit in Figure 7–2 if R2 and R4 were doubled?
18. If a $3\frac{1}{2}$-inch-thick beam is covered with 3/4-inch plywood on one side and 5/8-inch drywall on the other side, what is the total thickness of the assembly?
19. If 3/4 gallon of paint is taken from a 5-gallon pail in the morning and $1\frac{3}{8}$ gallons are taken in the afternoon, how much paint remains in the pail?
20. According to the information shown in Figure 7–3, how many hours did the carpenters work?
21. How many hours did all crews in Figure 7–3 work together?
22. In Figure 7–3, how many more hours did the carpenters work than the electricians worked?
23. If electrician A uses $27\frac{2}{3}$ feet and $45\frac{3}{4}$ feet from a spool of cable that had $221\frac{1}{2}$ feet and electrician B uses $37\frac{1}{6}$ feet from the same spool, how much cable is left on the spool?
24. A motor brush is $1\frac{5}{8}$ inches long. How long is it after 13/16 wears away?
25. How much needs to be removed from a board $7\frac{7}{16}$ inches wide to make a board $5\frac{3}{8}$ inches wide?

Multiplying Fractions

To multiply two or more fractions, multiply the numerators. Multiply the denominators. Write as a fraction with the product of the numerators over the product of the denominators. Express the answer in the lowest terms.

EXAMPLE 10 Multiply 1/2 × 4/5.

1/2 × 4/5 = 4/10 × 2/5

To multiply any combination of fractions, mixed numbers, and whole numbers, write the mixed numbers as fractions. Write the whole numbers as fractions with a denominator of 1. Multiply the numerators. Multiply the denominators. Express the answer in its lowest terms.

EXAMPLE 11 Multiply $2\frac{1}{3}$ × 4 × 4/5

1. Write the mixed number $2\frac{1}{3}$ as the fraction 7/3.
2. Write the whole number 4 as a fraction with a denominator of 1. (4/1)
3. Multiply the numerators. 7 × 4 × 4 = 112
4. Multiply the denominators. 3 × 1 × 5 = 15
5. Express the answer in lowest terms. 112/15 = $7\frac{7}{15}$

Problems

Multiply the following quantities:

26. 7/16 × 3/5
27. 7/8 × 2/3
28. $2\frac{3}{4}$ × 4/5
29. $5\frac{9}{16}$ × 3 × 2/7
30. $3\frac{1}{2}$ × $4\frac{3}{4}$ × $7\frac{1}{3}$

Dividing Fractions

Division is the inverse of multiplication. Dividing by 4 is the same as multiplying by 1/4. So 4 is the inverse of 1/4, and 1/4 is the inverse of 4. (Remember, 4 can be written as 4/1.) The inverse of 5/16 is 16/5. To divide fractions, invert the divisor (the part the other number is being divided by) and multiply. Express the answer in lowest terms.

EXAMPLE 12 Divide 7/8 by 2/3. 7/8 ÷ 2/3

1. Invert the divisor. The inverse of 2/3 is 3/2.
2. Multiply. 3/2 × 7/8 = 21/16
3. Express in lowest terms. 21/16 = $1\frac{5}{16}$

	Carpenters	Electricians	Plumbers	Painters
Monday	27½ hours	15¼ hours	12⅚ hours	14⅛ hours
Tuesday	25¾ hours	16⅔ hours	12⅙ hours	13⅞ hours
Wednesday	26⅞ hours	15½ hours	13¼ hours	15⅓ hours
Thursday	12⅝ hours	7⅞ hours	11⅞ hours	7⅞ hours
Friday	18⅓ hours	3⅚ hours	3 hours	8¼ hours

Figure 7–3

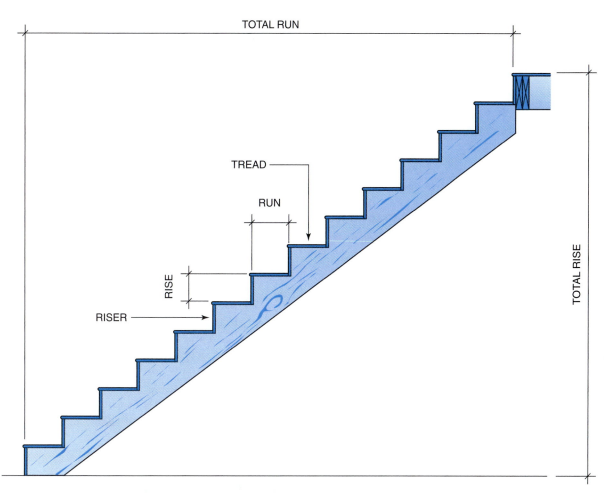

Figure 7–4

To divide any combination of fractions, mixed numbers, and whole numbers, write the mixed number as a fraction. Write the whole number over the denominator 1. Invert the divisor and multiply. Express the answer in lowest terms.

EXAMPLE **13** Divide 3/4 by by 2⅕. 3/4 ÷ 2⅕

1. Write the mixed number as a fraction. 2⅕ = 11/5
2. Invert the divisor. 11/5 inverted is 5/11.
3. Multiply. 3/4 × 5/11 = 15/44

Problems

Divide the following quantities:

31. 3/4 ÷ 2/3   32. 4/5 ÷ 1/8   33. $2^{15}/_{16}$ ÷ 2/3

34. 9/16 ÷ $3^7/_{12}$ 35. $4^5/_8$ ÷ $7^2/_3$

Solve the following problems:

36. How many air-conditioning systems can be charged with a cylinder containing 24½ pounds of refrigerant if each system takes 1¾ pounds?

37. The stair shown in Figure 7–4 has 14 risers. What is the total rise of the stair if each riser is 7⅜ inches?

38. What is the total run of the stair shown in Figure 7–4 if each tread is 9⅝ inches?

39. The plenum in Figure 7–5 is fastened to a furnace with equally spaced screws. What is the distance between the centers of the first and last screws?

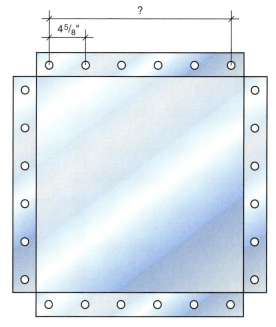

Figure 7–5

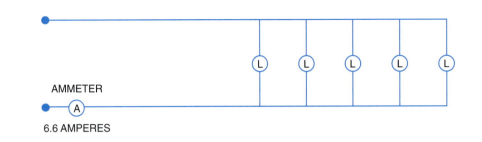

AMMETER

A

6.6 AMPERES

Figure 7-6

40. What is the cost of 1,455 feet of cable at $0.16 per foot?

41. The lamps in Figure 7-6 draw a total of 6.6 amperes. Assuming all the lamps are the same and draw the same amount of current, how many amperes will 3 lamps draw?

42. A piece of duct measures 24 inches. When 2 pieces are fitted together, $1\frac{1}{2}$ inches is allowed on each end for joining the pieces. If $25\frac{1}{2}$ pieces are used on a job, what is the total length of the assembled duct?

43. Find the width in inches of each opening in the ceiling register shown in Figure 7-7.

44. Find the height in inches of each opening in the ceiling register shown in Figure 7-7.

45. The landing in a stair is 59 inches high, and each riser is $7\frac{3}{8}$ inches high. How many risers does it take to reach the landing? (Refer to Figure 7-4 for the definition of a riser.)

46. Three laborers work part time for 5 days. Laborer A works $2\frac{1}{2}$ hours every day. Laborer B works $3\frac{3}{4}$ hours every day. Laborer C works $3\frac{1}{3}$ hours every day. They are paid $14.85 per hour. What is the total pay for these three laborers?

47. If laborer B in question 46 were replaced by a skilled worker at $19.40 per hour, how much difference would there be in the total pay for the week?

48. A carpenter and helper can place 9 floor joists in 1/4 hour. How long will it take them to place 126 floor joists?

49. An electrical contractor bills a customer $62.00 per hour for $4\frac{3}{4}$ hours per day for 5 days. He pays his workers 1/3 of what he bills, and he pays 1/4 of the bill for materials. How much money is left?

50. A wiring job calls for 32 pieces of 1/2-inch conduit $7\frac{1}{2}$ feet long, 8 pieces $1\frac{3}{4}$ feet long, and 3 pieces $7\frac{2}{3}$ feet long. If 1/5 is allowed for waste, how many feet of conduit are needed?

Decimal and Common Fraction Equivalents

To change a common fraction to an equivalent decimal, divide the numerator by the denominator.

EXAMPLE 14 Express 5/8 as a decimal.

1. Write 5/8 as a division problem. $8\sqrt{5}$
2. Place a decimal point after the 5 and add zeros as necessary. $8\sqrt{5.000}$
3. Place the decimal point in the answer directly above the decimal point in the dividend. $\dfrac{0.625}{8\sqrt{5.000}}$

To express a decimal as a common fraction, write the number after the decimal point as the numerator of a common fraction. Write the denominator as 1 followed by as many zeros as there are digits to the right of the decimal point. Express the common fraction in lowest terms.

EXAMPLE 15 Express 0.125 as a common fraction.

1. Write 125 as the numerator.
2. Write the denominator as 1 followed by three zeros. 125/1,000
3. Express the answer in lowest terms. 125/1,000 = 1/8

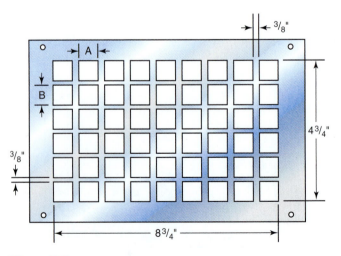

Figure 7-7

Problems

Express the following common fractions as decimal equivalents:

51. 3/8

52. 3/4

53. 1/4

54. 5/9

55. 27/32   (Round off to three places.)

Express the following decimals as common fraction equivalents:

56. 0.78125

57. 0.625

58. 0.6875

59. 0.5312

60. 0.546875

Solve the following problems:

61. In setting forms for a sidewalk, a carpenter is asked to raise the form 3/4 inch above an elevation of 141.35 feet. What is the new elevation?

62. A bearing with an inside diameter of $1\frac{1}{4}$ inches is found to be 0.008 oversize for the armature shaft. In decimal inches, what should the diameter be to fit the shaft? Allow 0.002 clearance for lubrication.

63. The slight sideward motion of a shaft is called end play. The end play of a motor shaft should not be more than 1/32 inch. One motor has end play of 0.0305 inch. Is this end play more than 1/32?

64. The elevation of a foundation is 19.60 inches. A floor frame is built on top of the foundation, adding another $13\frac{3}{4}$ inches. What is the elevation at the top of the floor frame in decimal inches?

65. The inside diameter of a tube is $1\frac{1}{4}$ inches. The wall of the tube is 0.15625 inch thick. What is the outside diameter in inches and fractional parts of an inch?

Chapter 8 | Linear Measure

Glossary of Terms

metric system a system of measurement based on 10 (often called the SI system).

nominal dimensions the dimensions of a product before allowances or adjustments are made. The sizes of many construction materials are identified by their nominal dimensions. For example, nominal dimensions of lumber are the dimensions of lumber before it is dried and planed or the dimensions of masonry units including the mortar joints.

perimeter the distance around the outside of a shape.

scale the system of graduating a measuring device, such as a ruler, into units and fractional parts of those units.

U.S. customary system the system of measurements we use in America based on inches, feet, quarts, gallons, pounds, and so on.

The ability to measure distances and to work mathematically with linear measure is fundamental to construction. Nearly every piece that is assembled into a house must be measured to fit. Major assemblies are made up of many small parts, and so it is necessary to be able to plan the size of those assemblies. To estimate the cost of building a house, it is vital to be able to perform mathematical operations with the linear measure of the parts.

Reading a U.S. Customary Scale

Most measurement of distances in the U.S. construction industry is done in feet and inches—units of measure in the **U.S. customary system.** It is nearly impossible to work in the building trades without being able to read a foot-and-inch scale, such as is found on a tape measure or ruler. A **scale** is the system of graduating a measuring device, such as a ruler, into units and fractional parts of those units. The major units of measure are the yard, foot, inch, and fractions of an inch (Fig. 8–1). There are 3 feet in a yard and 12 inches in a foot. Each inch can be divided into fractional parts of an inch. If an inch is divided in two, the parts are halves of an inch. Each of the two halves can be divided in two, making four quarters of an inch. If the quarters are divided in two, the parts are eighths. If the eighths are divided in two, the parts are sixteenths of an inch.

The marks on a U.S. customary scale are called graduations. If the scale is graduated in sixteenths, there are five sizes of graduation marks. The whole-inch marks are the longest; the half-inch marks are the next longest; quarter

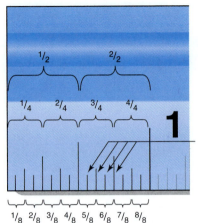

THE SHORTEST MARKS AND EVERY MARK THAT IS LONGER REPRESENTS SIXTEENTHS.

Figure 8–2

inch is next; then there are the eighths, and the smallest are the sixteenth-inch marks (Fig. 8–2). To read the value of a fractional inch graduation, first see what the length of the mark is to determine if it is halves, quarters, eighths, sixteenths, and so on. Then count the number of marks that are that big or bigger from the last whole-inch mark (Fig. 8–3).

Problems

Write the dimensions shown by the circled numbers in Figure 8–4.

1. _____
2. _____
3. _____
4. _____
5. _____

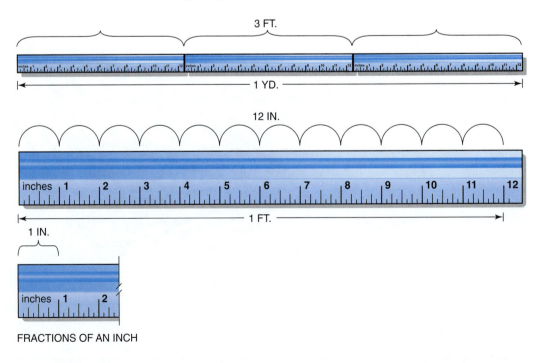

FRACTIONS OF AN INCH

Figure 8–1

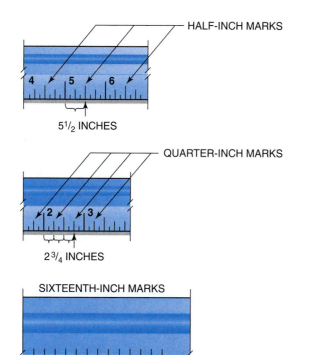

HALF-INCH MARKS

$5\frac{1}{2}$ INCHES

QUARTER-INCH MARKS

$2\frac{3}{4}$ INCHES

SIXTEENTH-INCH MARKS

3　　　**4**

$3\frac{5}{16}$ INCHES

Figure 8–3

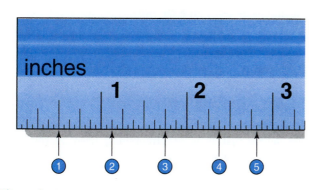

inches

1　　**2**　　**3**

①　②　③　④　⑤

Figure 8–4

Nominal Dimensions

Many things in construction are sized by their **nominal dimensions.** Those are the dimensions of the product before any allowances or adjustments are made. A typical concrete block is 8 inches × 8 inches × 16 inches nominal. The actual dimensions of that block are $7\frac{5}{8}$ inches × $7\frac{5}{8}$ inches × $15\frac{5}{8}$ inches, leaving room for 3/8-inch mortar joints. The following are the nominal dimensions of some common lumber sizes:

Nominal	Actual
1×4	$3/4 \times 3\frac{1}{2}$
1×6	$3/4 \times 5\frac{1}{2}$
2×4	$1\frac{1}{2} \times 3\frac{1}{2}$
2×6	$1\frac{1}{2} \times 5\frac{1}{2}$
2×8	$1\frac{1}{2} \times 7\frac{1}{4}$
2×10	$1\frac{1}{2} \times 9\frac{1}{4}$

Combining Feet and Inches

It is frequently necessary to work with dimensions of feet and inches combined. To combine them, start by converting everything to inches. Multiply the number of feet by 12 and add the inches.

EXAMPLE ❶ Express 5 feet 4 inches as inches.

Multiply the number of feet by 12. $5 \times 12 = 60$
Add the inches. $60 + 4 = 64$ in.

When your calculations are finished, you should express the answer in terms of feet and inches. To convert inches to feet and inches, divide by 12 to find the number of whole feet. The remainder is the number of inches in addition to the whole feet.

EXAMPLE ❷ Express 131 inches as feet and inches.

Divide the number of inches by 12.

$$\begin{array}{r} 10 \text{ R}11 \\ 12\overline{)131} \\ \underline{120} \\ 11 \end{array}$$

131 in. = 10 ft. 11 in.

Problems

Convert the following to inches:

6. 1 ft. 4 in.

7. 15 ft. 7 in.

8. 3 ft. 1 in.

9. 3 ft. 6 in.

10. 21 ft. 9 in.

Convert the following to feet and inches:

11. 15 in.

12. 74 in.

13. 71 in.

14. 125 in.

15. 23 in.

Add or subtract the following and express the result in feet and inches:

16. 2 ft. 6 in. + 4 ft. 3 in.

17. 9 ft. 7 in. + 10 ft. 8 in.

18. 12 ft. 3 in. − 1 ft. 2 in.

19. 3 ft. 4 in. − 2 ft. 5 in.

20. 14 ft. 5 in. − 3 ft. 6 in.

Solve the following problems:

21. What is the width of the wall space without the door in Figure 8–5?

22. A piece of PVC pipe 2 feet 3 inches long is cut from a 10-foot 0-inch piece. How much pipe is left?

23. A furnace flue is made from 4 pieces of pipe. After they are fitted together, each piece of flue pipe is 1 foot 10½ inches long. What is the total length of the flue?

24. A wiring job requires 43 feet 3 inches of Electrical Metallic Tubing (EMT). If the EMT comes in 10-foot lengths, how many pieces are needed, and how much must be cut off the last piece?

25. The control wire for a gas burner runs 6 inches, 2 feet 4 inches, 6 feet 9 inches, 12 feet 2 inches, and 10 inches. What is the total length of the control wire?

26. In Figure 8–6, how many feet and inches is it from the left end of the building to the centerline of the window?

27. In Figure 8–6, what is the width of bedroom 1?

28. In Figure 8–6, what is the total width of the outside of the house in feet and inches?

29. A duct must be 9 feet 5½ inches long. When the sections of duct are assembled, they are each 1 foot 10½ inches long. How many pieces are needed, and how much must be cut off the last piece?

30. Support straps for a duct are each 1 foot 2 inches long. How many straps can be cut from a 50-foot roll of metal?

Metric System

The **metric system** is a decimal system. That is, it is based on dividing units of measure by 10 or multiplying them by 10. The base unit for linear measure in the metric system is the *meter*. A meter is slightly longer than your leg. To work with metric units you need to know the most common metric prefixes.

For units larger than a meter:

10 meters = 1 *deka*meter (about the width of a small house)

10 dekameters or 100 meters = 1 *hecto*meter (about the length of a football field)

10 hectometers or 1,000 meters = 1 *kilo*meter (a little more than one-half mile)

For units smaller than a meter:

One-tenth of a meter = 1 *deci*meter (the length of a child's crayon)

One-tenth of a decimeter or one one-hundredth of a meter = 1 *centi*meter (the thickness of a pencil)

One-tenth of a centimeter or one one-thousandth of a meter = 1 *milli*meter (about the thickness of a dime)

The metric system is not frequently used by the construction industry in the United States, but some materials and parts are built for the world market and use metric dimensions.

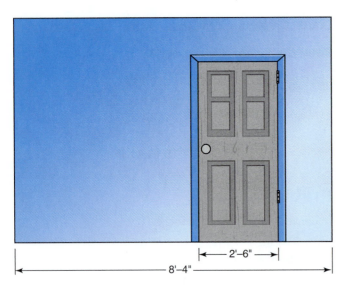

|←— 2'-6" —→|

|←——————— 8'-4" ———————→|

Figure 8–5

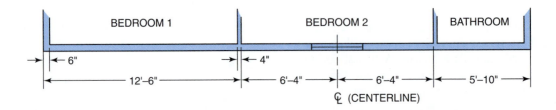

| BEDROOM 1 | BEDROOM 2 | BATHROOM |

←| 6" | ←| 4" |

|←——————— 12'-6" ———————→| |←— 6'-4" —→|←— 6'-4" —→|←— 5'-10" —→|

℄ (CENTERLINE)

Figure 8–6

Problems

Match the item from the left column with the metric dimension in the right column.

Item	Metric Dimension
31. Thickness of plywood	20 mm
32. New coil of cable	100 M
33. Uncut length of plastic pipe	3 M
34. Length of a large housing development	1.1 km
35. Width of a wood beam	22 cm

Perimeter Measure

The distance around the outside of something is its **perimeter.** We could find the perimeter of a square or a rectangle by adding the lengths of all four sides. A rectangle is a four-sided shape in which the opposite sides are of equal length (Fig. 8–7). The perimeter (*P*) of a rectangle can be found by multiplying 2 × the length (*l*) and 2 × the width (*w*) and adding the two products together. This formula is written as $P = 2l + 2w$. Since a square is a rectangle with four equal sides, you can find the perimeter by multiplying 4 × the length of one side (*s*). This is written as $P = 4s$.

The perimeter of a circle is called the circumference, known as *C* (Fig. 8–8). The diameter (*D*) of a circle is the distance across a circle through its center. To find the circumference of a circle, we use a certain number represented by the Greek letter π, pronounced "pie." π is equal to 3.1416, or 22 ÷ 7. The circumference of a circle is found by multiplying π × the diameter of the circle. This formula is written as $C = \pi D$.

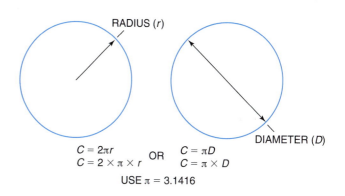

$$C = 2\pi r \qquad C = \pi D$$
$$C = 2 \times \pi \times r \quad \text{OR} \quad C = \pi \times D$$
$$\text{USE } \pi = 3.1416$$

Figure 8–8 Circumference of a circle.

Problems

Solve the following problems:

36. What is the perimeter of a house that is 28 feet wide and 40 feet long?

37. What is the perimeter of a square room that is 14 feet on each side?

38. How wide must insulation be to wrap around an 8-inch square duct?

39. How wide must insulation be to wrap around an 8-inch-diameter round duct?

40. What is the perimeter of a rectangular building lot that is 80 feet wide and 140 feet deep?

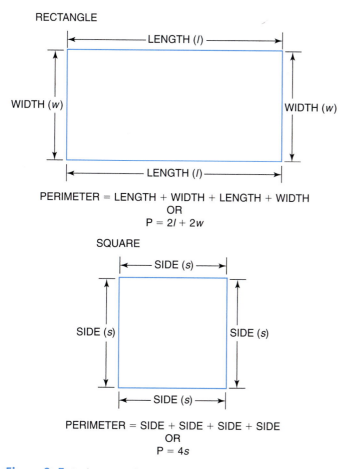

RECTANGLE

PERIMETER = LENGTH + WIDTH + LENGTH + WIDTH
OR
P = 2*l* + 2*w*

SQUARE

PERIMETER = SIDE + SIDE + SIDE + SIDE
OR
P = 4*s*

Figure 8–7 Perimeters of squares and rectangles.

Chapter 9 | Percent (Percentage)

OBJECTIVES

After completing this chapter, the student should be able to:

- find any percentage of any given number.
- determine a part's percentage of a whole.
- calculate the total amount of money owed when given the principal, rate, and time.
- solve basic problems involving simple interest.

Glossary of Terms

percentage (percent) hundredths of the whole.

rate percentage at which interest is calculated.

principal the amount of money invested or borrowed.

interest payment for the use of money; expressed as a percentage of the principal. Interest is usually stated.

amount principal plus interest (this is the total amount owed after borrowing).

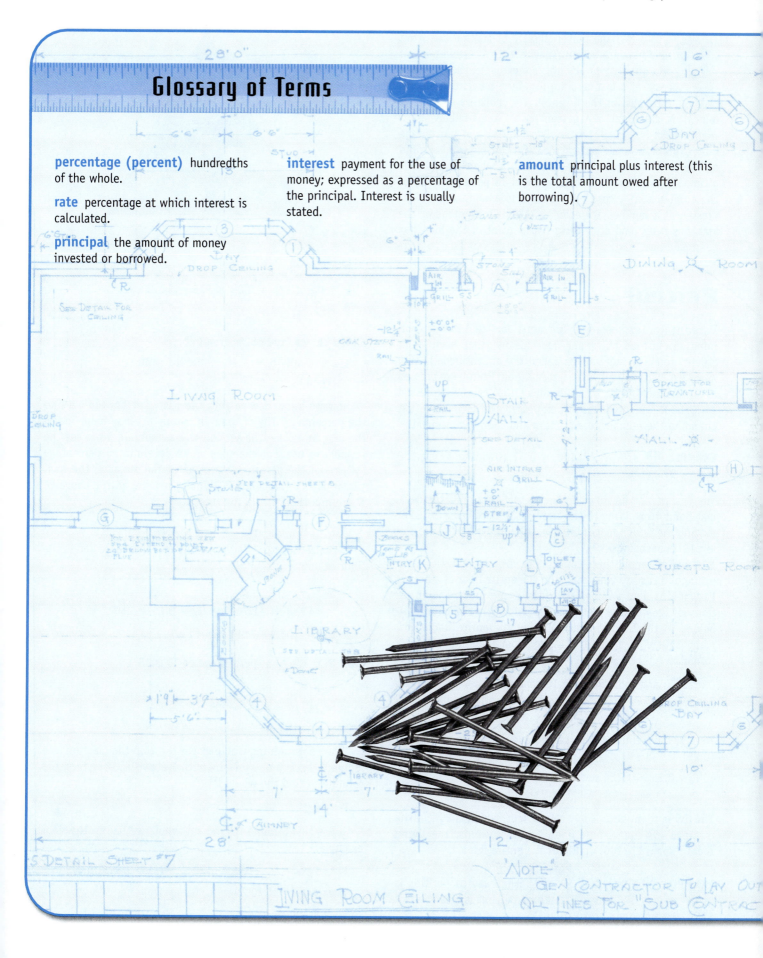

Percent (%) and percentage are terms that are often seen in construction. They refer to a certain portion of a quantity or amount. For example, the overhead costs of a business are usually a percentage of the total cost of the business. The moisture content of lumber is stated as the percentage of the lumber's total weight that is actually water. Percent (%) and percentage both express the same general idea, but the two words should not be used interchangeably. Use percent when a number is given. For example, "The concrete was 18 percent (18%) of the total material cost." Percentage is a noun and is expressed as a number. For example, "What percentage of the loan has been paid out? Forty-five percent of the loan has been paid out."

Percent

Percentage (expressed as "**percent**" or "**%**") means each hundred, just as miles per hour means miles each hour. ("Cent" refers to hundred, as in century or centipede.) Forty miles per hour (40 miles driven in 1 hour) can be expressed as the fraction 40/1; 25% (25 out of 100) can be expressed as 25/100. Percentage may be expressed as a decimal by first changing the number to a fraction and then to a decimal.

EXAMPLE 37% = 37/100 = 0.37 (Dividing 37 by 100 gives 0.37.)

The % sign may be considered a magnet that pulls the decimal point two places to the right (0.37 becomes 37%). Removing the % sign moves the decimal point two places to the left (37% becomes 0.37).

Now, let's apply these concepts to construction. Here's an example: Let's say you are laying the entire floor of a house. Part of this floor will belong to the living room. The entire floor is 1,764 square feet, and the part of the floor that will belong to the living room is 25% of the entire floor. How many square feet will the living room floor be? Use this formula:

$$\text{Part (living room floor)} = \text{Percentage (\%)} \times \text{Whole (entire floor)}$$
$$P = \% \times W$$

To multiply, convert the percentage (25%) to a decimal (0.25).

$P = 0.25 \times 1{,}764$ sq. ft.
$P = 441$ sq. ft.

The living room floor is 441 square feet, which is 25% of the entire 1,764-square-foot floor. In other words, the living room floor is 25/100, or 1/4 of the entire floor.

EXAMPLE ❷ Find 3% of $75. (Another way to say this is "What part of $75 is 3%?" or "Three percent is what part of $75?" We are solving for Part.)

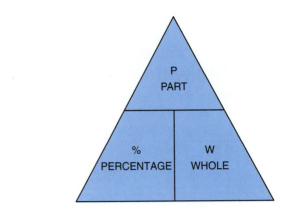

Figure 9–1

$P = \% \times W$
$P = 3\%$ of $75 = 0.03 \times \$75 = \$2.25$
$P = \$2.25$

EXAMPLE ❸ Find 150% of 400 lb.

150% of 400 lb. = 1.50 × 400 lb. = 600 lb.

$P = 600$ lb.

Now we know how to find the part, but how do we find the percentage? The diagram in Figure 9–1 may help. To use the diagram, cover the section you are solving for to see the formula to use. For example, when determining percentage (%), cover the percentage section of the diagram; the fractional parts P (Part) and W (Whole) remain, with P above W. Therefore, the formula for percentage is

$\% = P/W$

Let's use this formula to find the percentage we had in the floor example.

$\% = P/W$
$\% = 441$ sq. ft./1,764 sq. ft. = 0.25 = 25%
$\% = 25$ (The percentage is 25%.)

EXAMPLE ❹ What percentage of 75 is 15?

$\% = P/W$
$\% = 15/75 = 0.20 = 20\%$
$\% = 20$ (The percentage is 20%.)

When determining whole, cover the W in the diagram (see Figure 9–1); P and % remain, with P above %. Therefore, the formula to find the whole is

$W = P/\%$

Let's use this formula to find the whole we had in the floor example.

$W = P/\%$
$W = 441$ sq. ft./0.25 = 1,764 sq. ft.
$W = 1{,}764$ sq. ft.

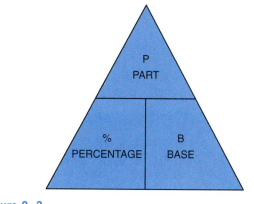

Figure 9–2

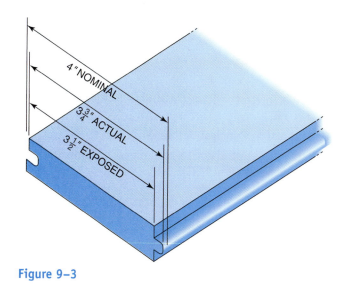

Figure 9–3

EXAMPLE ⑤ Three is 25 percent of what number? (Another way to say this is "Three is 25% of what whole?" We are solving for Whole.)

W = P/%

W = 3/0.25 = 12

W = 12

In actual usage, Whole (W) is sometimes called Base (B). The formula does not change; only the label changes. See Figure 9–2.

Problems

Solve the following problems:

1. P = 493; Percentage = 85%; B = ?

2. P = ?; Percentage = 55%; B = 125

3. P = $588; Percentage = ?; B = $840

4. It is estimated that 2,550 square feet of decking is needed for a roof. If an additional 20% is added for waste, what is the total amount required?

5. The total area to be covered with OSB is 1,250 square feet. How much should be ordered if 8% is added for cutting and waste?

6. The flooring board shown in Figure 9–3 is sold as 4-inch flooring, but part of the 4 inches is lost in shrinkage as the board dries and in machining. Another portion is in the tongue, which does not show on the finished floor. What percentage of the 4 inches shows when the board is installed?

7. In constructing a building, 2% of the total cost is allowed for excavation. What is the total cost of the building if the excavation costs $2,825?

8. In making mortar, lime is used in an amount equal to 12% of the cement. How much lime is necessary if 995 pounds of cement are used?

9. If 65 square feet of copper flashing are needed for a particular job and 20% is to be added for waste, how much should be ordered?

10. The total cost of building a house is $86,000. What percentage is spent on excavation, foundation, and framing if the total for these costs is $52,275?

11. A dealer buys an item at a cost of 6,600 and sells it for a profit of 12% of the cost. Find the amount of profit and the selling price. (Note: selling price = cost + profit.)

12. A contractor makes $18,000 profit on a job. If the profit is 12% of the total, what does the owner pay?

13. The grading on a $145,500 job comes to $2,182.50. What percentage of the total cost is this?

14. A project has actual costs of $3,464.00. If the contractor adds 15% for overhead and 18% of the cost for profit, what will the total charged to the customer be?

15. A contractor submits a bid for a building. The estimated cost is $181,693.21. Add to this 23% of the estimated cost (for the overhead) and 15% of the cost plus the overhead (for profit). What is the final bid?

Interest

When a person borrows money, the person lending the money usually charges the borrower a fee for the use of the money. An example is the fee that a credit card company charges when their card is used to buy an item. This fee is in addition to the price of the item. This fee is a form of **interest.** Banks also pay interest to depositors for the privilege of using the depositors' money. A builder often borrows money to build a house. Interest must be paid on the amount borrowed (the "**principal**") until the house is sold and the loan is repaid. Interest is usually expressed as a percentage of the principal or amount borrowed. Another term used for the interest percentage is **rate.**

The following formulas and examples will use the terms and definitions given at the beginning of this chapter.

Formulas

1. Simple interest = Principal × Rate × Time in years
   (I = PRT)

2. **Amount** = Principal + Interest

3. Rate = Interest divided by the product of Principal × Time in years (R = I/PT)

EXAMPLE 6 Find the simple interest and amount on $800 at 12% after 2 years.

Interest = Principal × Rate × Time in years
Interest = $800 × 12% × 2
　　　　 = $800 × 0.12 × 2
　　　　 = $192
Amount = Principal + Interest
Amount = $800 + $192 = $992

EXAMPLE 7 The annual rate of interest on $12,000 is 8%. Find the interest after a period of 3 months. (Hint: 3 months = 3/12 year = 1/4 year)

Interest = Principal × Rate × Time
Interest = $12,000 × 0.08 × 1/4
Interest = $240.00

Problems

Determine the simple interest and amount under each of the following conditions:

Principal	Rate of Interest	Time	Interest	Amount
16. $300	9%	4 years	_____	_____
17. $640	11%	1/2 year	_____	_____
18. $4,280	8%	3 months	_____	_____
19. $20,000	8½%	1 year	_____	_____
20. $1,080	9½%	1½ years	_____	_____

Solve the following problems:

21. A person pays 7% simple interest on a $140,000 construction loan. What is the interest paid in the first year of the loan?

22. A builder buys equipment on credit for $44,250 and the interest rate is 9% per year. What is the interest payment for the first year?

23. A builder develops a tract of 20 houses at a cost of $186,000 each. If a loan is made for 1/2 the cost of the houses at 6¾% simple interest, what interest is paid in the first year?

24. Two business partners have a bank balance of $57,942.00. After the first year, the interest brings their balance to $60,839.10. What rate of interest did they earn?

25. A carpenter borrows $10,000 at 6% interest to be repaid in annual payments of $2,000. What is the principal balance after the first payment is made? (Hint: Any part of the payment that is not used to pay interest is used to reduce the principal.)

The above problems only involved applying a simple interest rate to the original principal amount of a loan. It is usually customary that the principal is reduced as the loan is repaid. Each payment pays whatever interest is due at that time, then any remaining part of the payment is used to reduce the principal.

EXAMPLE 8 The Principal is $150,000 and the Interest to be paid in the first year is $11,250. What is the rate?

Rate = Interest ÷ (Principal × Time)
Rate = 11,250 ÷ (150,000 × 1)
Rate = 11,250 ÷ (150,000)
Rate = 0.075 or 7.5%

Problems

26. If a plumbing contractor borrows $12,000 on a short term loan of 6 months and the total amount he will pay is $12,540, what is the rate of interest on the loan?

27. What is the rate of interest paid on a loan of $15,000 if the loan is paid of in 24 payments of $678.12? Round off your answer to the nearest tenth of a percent.

Chapter 10 Area and Volume

OBJECTIVES

After completing this chapter, the student should be able to:

- ✪ determine the area of squares and rectangles.
- ✪ determine the area of triangles.
- ✪ determine the area of circles.
- ✪ determine the volume of rectangular solids and cubes.

Glossary of Terms

area the space inside a shape.

base of a triangle the side opposite the corner from which the height is measured. This can be any side of the triangle.

circle a shape in which every point on the perimeter is the same distance from a center point.

cube a three-dimensional shape in which height, width, and depth are all equal.

height of a triangle the length of a line drawn perpendicular to one side of a triangle and extending to the opposite corner.

rectangle a shape with four 90° corners and with opposite sides of equal length.

solid a three-dimensional shape. Spheres and cubes are two types of solids.

square a shape with four 90° corners and four equal-length sides. Square also refers to the result of multiplying a number by itself.

triangle a shape formed by three sides.

volume the space enclosed by a three-dimensional figure.

n Chapter 8, we discussed squares, rectangles, and circles, and in this chapter, we shall talk about them again. We will also focus on another shape you will encounter on the job: the **triangle,** a shape formed by three sides. Triangles are found throughout construction. A common example is the gable of a house. A gable is the triangle formed by two sloping rafters at one end of a roof and a straight line connecting the two rafter bottoms. Electricians also work with triangles sometimes when calculating values in alternating current.

Area of Squares and Rectangles

The **area** of a shape is the size of the inside of the shape. Area is always measured in square units—for instance, square inches, square feet, and square yards. One inch times one inch equals one square inch, one foot times one foot equals one square foot, and so on (Fig. 10–1).

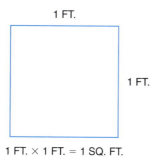

1 FT.

1 FT.

1 FT. × 1 FT. = 1 SQ. FT.

Figure 10–1

A **rectangle** is a shape with four 90° corners and with opposite sides of equal length. To find the area of a rectangle, multiply the width of the rectangle by its length. (All dimensions must be in the same units. For example, inches cannot be multiplied by feet, nor can yards be multiplied by meters.) A **square** is a shape with four 90° corners and equal-length sides.

EXAMPLE ❶ Find the area of a rectangle 18 inches wide and 3 feet long.

1. Change the 18 inches to feet.   18 in. = 1½ ft.
2. Multiply the width by the length.   1½ ft. × 3 ft. = 4½ sq. ft.

This area can also be computed by changing 3 feet to 36 inches, but the answer would be 648 square inches, which would have to be divided by 144 to find square feet.

U.S. Customary Units of Square Measure
144 square inches = 1 square foot
9 square feet = 1 square yard
4,840 square yards = 1 acre

Figure 10–2

Metric Units of Square Measure
100 decimeters squared (100 dm²) = 1 square meter (1 M²)
100 centimeters squared (100 cm²) = 1 square decimeter (1 dm²)
100 millimeters squared (1 mm²) = 1 square centimeter (1 cm²)
1,000,000 millimeters squared = 1 square meter (1 M²)

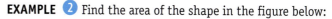

Figure 10–3

In the U.S. Customary System, area is measured in square inches, square feet, square yards, and square miles (Fig. 10–2).

In the metric system, the base unit of square measure is the square meter (Figure 10–3).

In construction, it is often necessary to find the areas of shapes that are made up of combinations of squares, rectangles, and other shapes. To find these areas, break the shape into its individual parts, compute the area of each part, and then add the separate areas to find the total.

EXAMPLE ❷ Find the area of the shape in the figure below:

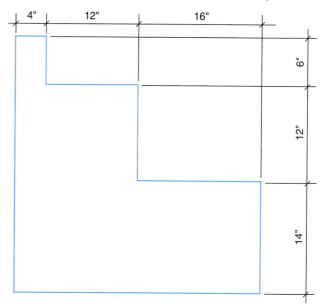

This shape is actually made up of three rectangles.

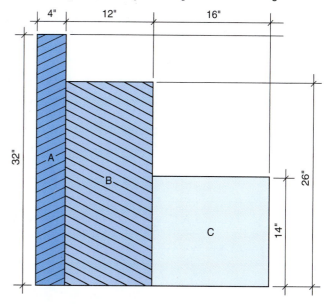

Find the dimensions of each of the three rectangles.
4 in. × 32 in., 12 in. × 26 in., and 16 in. × 14 in.

Calculate the area of each part. 4 in. × 32 in. =
128 sq. in., 12 in. × 26 in. = 312 sq. in., 16 in. ×
14 in. = 224 sq. in.

Add the areas to find the total. 128 sq. in. + 312 sq. in.
+ 224 sq. in. = 760 sq. in.

Problems

Find the areas of the following squares and rectangles:

1. Square with 11-inch sides.

2. Rectangle 9 feet by 14 feet.

3. Square with sides 6 feet 0 inches.

4. Rectangle with sides 10 feet 0 inches by 12 feet
   6 inches. Express the answer in square feet.

5. Rectangle with sides 3 feet 6 inches by 14 feet
   3 inches. Express the answer in square feet and
   decimal fractions of a square foot.

Solve the following problems:

For problems 6 through 8, refer to Figure 10–4.

6. What is the area of the carport?

7. What is the area of the living room?

8. What is the area of the family room?

9. How many square feet of plywood are needed to cover
   the deck shown in Figure 10–5?

10. The grill shown in Figure 10–6 has equally spaced
    openings. What is the total area of the openings in
    the grill?

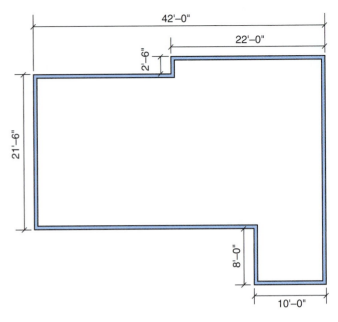

Figure 10–5

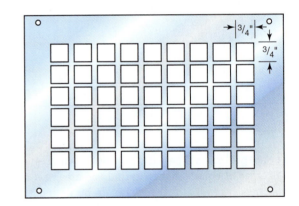

Figure 10–6

Figure 10–4

11. A particular type of lighting fixture can illuminate up
    to 80 square feet. How many fixtures would be required
    for a room 12 feet 6 inches by 18 feet 6 inches?

12. How much paint is required to paint the walls in two
    rooms at the rate of 400 square feet per gallon if one
    room is 14 feet by 12 feet and the other is 22 feet by
    16 feet? The walls in both rooms are 8 feet high.
    Round your answer off to the nearest half gallon.

13. The lighting for a game room requires 5 watts per
    square foot. How many watts are required if the room
    is 14 feet 9 inches wide by 20 feet 0 inches long?

14. What is the surface area of the drain field shown in
    Figure 10–7?

15. Three cold-air ducts are each 14 inches by 20 inches.
    What is the total area of the cold-air ducts?

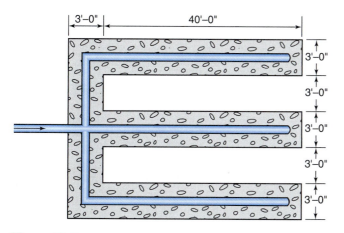

Figure 10–7

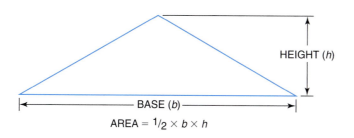

AREA = $\frac{1}{2} \times b \times h$

Figure 10–8

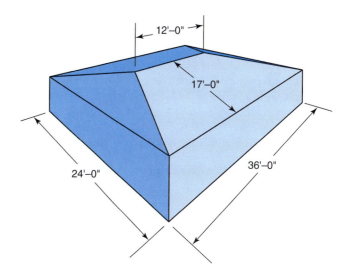

Figure 10–9

Area of Triangles

The **height** of a triangle is the length of a line drawn perpendicular to one side of the triangle and extending to the opposite corner. The **base** of a triangle is the side opposite the corner from which the height is measured. The area of a triangle is found by multiplying 1/2 × the length of the base (*b*) × the height (*h*) (Fig. 10-8).

EXAMPLE 🔵 Find the area of the triangle shown below.

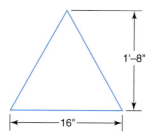

1. Convert 1 foot 8 inches to inches. 1 ft. 8 in. = 20 in.
2. Multiply 1/2 × 16 inches × 20 inches. 1/2 × 16 in. × 20 in. = 160 sq. in.

Problems

Solve the following problems:

16. What is the area of a triangle with a base of 8 feet and a height of 14 feet?

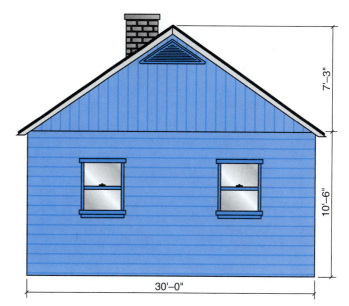

Figure 10–10

17. What is the area of a triangle with a base of 36 inches and a height of 14 inches?

18. What is the area of a building lot with 120 feet of road frontage and the two side property lines meeting 214 feet from the road?

19. What is the area of the roof shown in Figure 10-9?

20. What is the area of the end of the house shown in Figure 10-10?

Area of Circles

A **circle** is a shape in which every point on the perimeter is the same distance from a center point. The area of circular shapes is found by using the formula Area (*A*) = pi (π) × radius squared (*r*²). This is written as $A = \pi r^2$. Remember from Chapter 8 that π is a constant number used in working with circles and that the value of π is approximately 3.1416, or 22/7. The radius (*r*) of a circle is one-half its diameter.

EXAMPLE ④ How many square feet of surface area are there in a circle with a radius of 9 feet 6 inches?

$$A = \pi r^2$$

Area = 3.1416 × 9.5 ft. × 9.5 ft. = 283.53 sq. ft.

Find the areas of the following circular shapes.

21. Circle with a radius of 6 inches.

22. Circle with a radius of 3 feet 9 inches.

23. Circle with a diameter of 14 meters.

24. Circle with a diameter of 62 feet 6 inches.

25. Half circle with a radius of 18 inches.

Solve the following problems:

26. A conduit has an inside diameter of 2.875 inches. What is the cross-sectional area of the inside of the conduit (the area of the opening)?

27. A circular tabletop with a diameter of 36 inches is covered with plastic laminate. The laminate is cut from a piece 36 inches square. What is the area of the waste?

28. A semicircular bay is built to enclose one side of a spiral staircase, as shown in Figure 10–11. What is the area of the floor inside the bay?

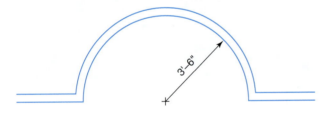

3'-6"

Figure 10–11

29. What is the difference in the cross-sectional area (area of the opening) between a 12-inch-diameter culvert pipe and an 18-inch-diameter culvert pipe?

30. An attic area requires 400 square inches of opening area for ventilation. How many 8-inch-diameter openings are needed?

Volume of Rectangular Solids, Cubes, and Cylinders

A **solid** is a three-dimensional shape. Spheres and cubes are two types of solids. **Volume** is the space enclosed by a three-dimensional figure. A rectangular solid is a three-

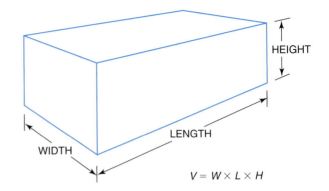

HEIGHT

LENGTH

WIDTH

$V = W \times L \times H$

Figure 10–12

dimensional figure with a rectangular base. The volume of a rectangular solid is found by multiplying the width of the base by the length of the base by the height. In other words, the volume is found by multiplying the area of the base by the height (Fig. 10–12). A **cube** is a solid in which the width, length, and height are all equal. Volume is measured in cubic units: cubic inches, cubic feet, cubic yards, and so forth.

EXAMPLE ⑤ Find the volume of a rectangular solid that is 4 inches wide, 2 feet long, and 3 inches high.

1. Convert the length into inches.   2 ft. = 24 in.
2. Multiply $w \times l \times h$.   4 × 24 × 3 = 288 cu. in.

It will be helpful to memorize the equivalents shown in Figure 10–13.

U.S. Customary Units of Cubic Measure
1 cubic yard = 27 cubic feet
1 cubic foot = 1,728 cubic inches

Figure 10–13

A cylinder is a solid shape that has a round cross section. To find the volume of a cylinder simply multiply the area of the end or cross section by the length.

Problems

Find the volumes of each of the following solids:

31. Cube with sides of 4 feet.

32. Rectangular solid with a width of 2 inches, length of 4 inches, and length of 10 inches.

33. Cube with sides of 10 feet 6 inches.

34. Rectangular solid with a width of 5½ inches, length of 3 feet 3 inches, and height of 8 inches.

35. Rectangular solid with a length of 4 feet, width of 7¼ inches, and height of 1/2 inch.

Solve the following problems:

36. What is the volume in cubic feet of a 10-foot 0-inch by 12-feet 6-inch room with an 8-foot ceiling?

37. A board foot is the amount of lumber contained in a board that is 1 inch thick, 1 foot wide, and 1 foot long (Fig. 10–14). How many board feet are contained in a board 1 inch thick by 6 inches wide by 12 feet long?

38. How many board feet are contained in ten 8-foot 2 × 4s? Note: A 2×4 is actually 1½ inches by 3½ inches. 2 × 4 represents the *nominal* dimensions of the piece; in figuring board feet, the nominal dimensions are used.

39. A cellar excavation needs to be 32 feet wide by 44 feet long by 6 feet deep. If the earth is hauled away from the site in dump trucks that carry 12 cubic yards each, how many truckloads must be removed? Round your answer to the nearest truckload.

40. What is the volume in cubic inches of an electrical box that measures 3 inches by 2 inches by 2½ inches?

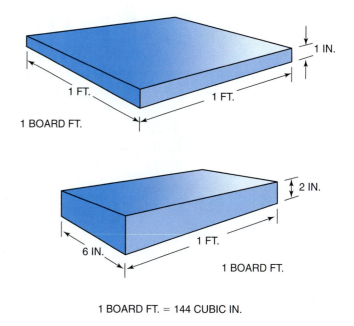

1 BOARD FT. = 144 CUBIC IN.

Figure 10–14 **1 board foot = 144 cubic inches.**

Chapter 11 | Right Angles

OBJECTIVES

After completing this chapter, the student should be able to:

- determine the hypotenuse of a right triangle if given the two adjacent sides.
- determine the length of one side of a right triangle if given the other side and the hypotenuse.

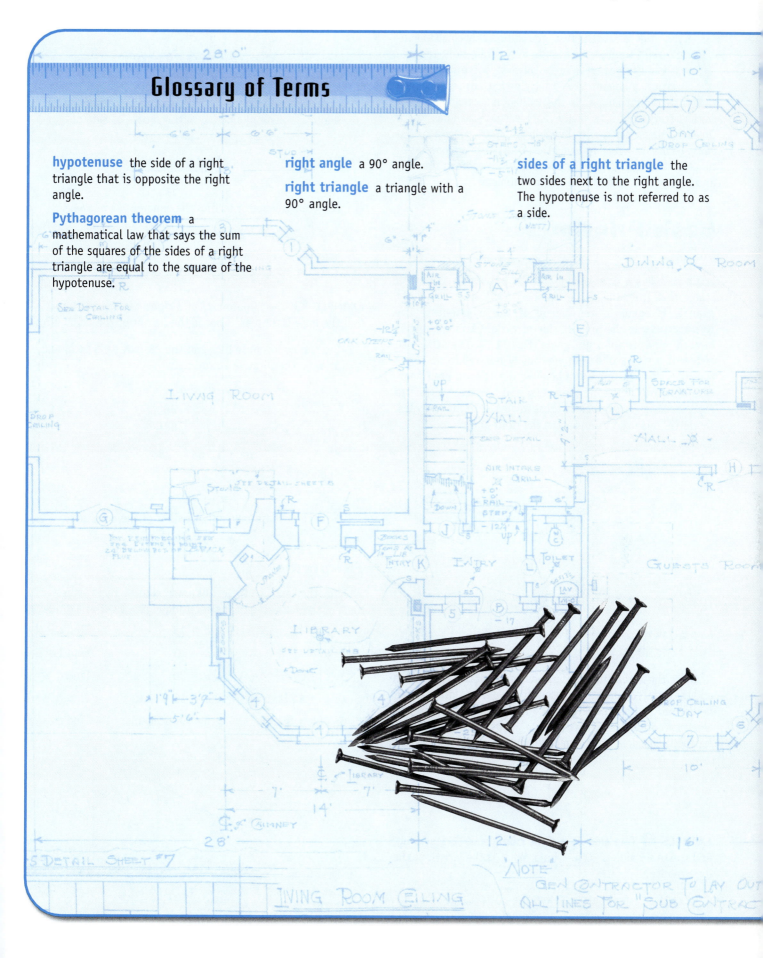

Glossary of Terms

hypotenuse the side of a right triangle that is opposite the right angle.

Pythagorean theorem a mathematical law that says the sum of the squares of the sides of a right triangle are equal to the square of the hypotenuse.

right angle a 90° angle.

right triangle a triangle with a 90° angle.

sides of a right triangle the two sides next to the right angle. The hypotenuse is not referred to as a side.

I f you work in construction, you will quickly come to consider right triangles as friends. Very often, right-triangle math is the easiest way to solve layout problems and electrical problems. The principle that is used so much with right triangles is the Pythagorean theorem. That sounds scary, but it is really fairly easy. Pythagoras was a Greek mathematician who lived about 3,000 years ago. Because he discovered and proved the principle, it is named after him. Most of this chapter is about the Pythagorean theorem and how you can apply it to solving problems.

Basic Principles

A **right angle** is a 90° angle, and a **right triangle** is a triangle that has a right angle. The **Pythagorean theorem** states that the sum of the squares of the sides of a right triangle is equal to the square of the hypotenuse. The **hypotenuse** is the longest side of a right triangle and is located opposite the right angle (Fig. 11–1). The other two **sides of a right triangle** are usually referred to simply as the sides. It is customary in labeling triangles to use uppercase letters to represent the angles and lowercase letters to represent the sides opposite those angles. In right triangles, the letter C is used to represent the right angle.

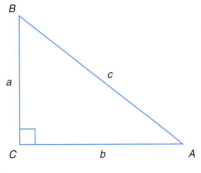

Figure 11–1

Using the letter c to represent the length of the hypotenuse and the letters a and b to represent the two sides, formulas can be derived to find the length of any side or the hypotenuse when two others are known using $a^2 + b^2 = c^2$. The formula for the Pythagorean theorem can be rearranged as follows:

$$c^2 = a^2 + b^2$$
$$a^2 = c^2 - b^2$$
$$b^2 = c^2 - a^2$$

EXAMPLE ❶ One side of a right triangle is 12 feet long, and the other side is 20 feet long. What is the length of the hypotenuse?

1. Let the 12-foot side be a and the 20-foot side be b, and arrange the formula accordingly.

$$c^2 = a^2 + b^2$$
$$c^2 = 12 \times 12 + 20 \times 20$$

2. Square side a and side b.

$$c^2 = 144 + 400$$

3. Add the square of side a and the square of side b.

$$c^2 = 544$$

4. Find the square root of both sides of the formula.

$$\sqrt{c^2} = c$$

Use your calculator to find the square root of 544.

$$\sqrt{544} = 23.3$$

5. Express your answer in feet.

$$c = 23.3 \text{ ft.}$$

EXAMPLE ❷ The hypotenuse of a right triangle is 42 feet, and the base is 28 feet. What is the height of the triangle?

1. Let the hypotenuse be c and the 28-foot base be b, and arrange the formula accordingly.

$$a^2 = c^2 - b^2$$
$$a^2 = 42^2 - 28^2$$

2. Square the hypotenuse and side b.

$$a^2 = 1,764 - 784$$

3. Subtract.

$$a^2 = 980$$

4. Find the square root of both sides of the equation.

$$a = 31.3 \text{ ft.}$$

Problems

Find the length of the unknown side or hypotenuse in each of the following triangles, using Figure 11–2 as a guide:

1. $a = 12$ $b = 15$ Find c
2. $a = 17$ $c = 30$ Find b
3. $c = 60$ $b = 30$ Find a
4. $a = 14$ ft. 6 in. $c = 16$ ft. 3 in.   Find b
5. $a = 4$ ft. 6 in. $b = 14$ ft. 0 in.   Find c

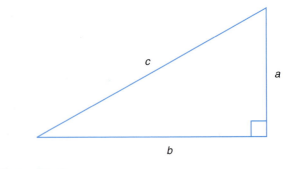

Figure 11–2

Solve the following problems:

6. What is the length of the rafters for the house shown in Figure 11–3?

7. If a 24-foot ladder is placed against a building, with the base of the ladder 6 feet from the building, at what height will the ladder touch the building?

8. One end of a guy wire is attached to an antenna tower at 40 feet from the ground, and the other end is attached to an anchor 14 feet 6 inches from the tower (Fig. 11–4). How long is the guy wire?

9. A 15-mile power line can be straightened out with a new right-of-way (Fig. 11–5). How long will the new power line be? (Hint: Break the figure into two right triangles.)

10. What is the length of pipe *x* in Figure 11–6? Do not make allowances for fittings.

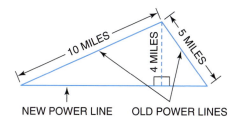

Figure 11–5

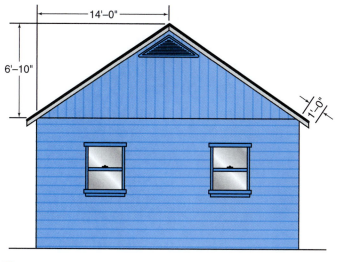

Figure 11–3

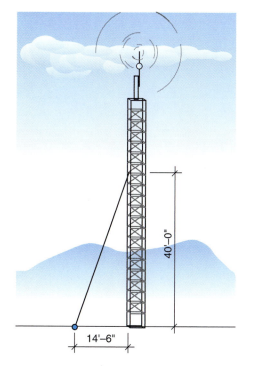

Figure 11–4

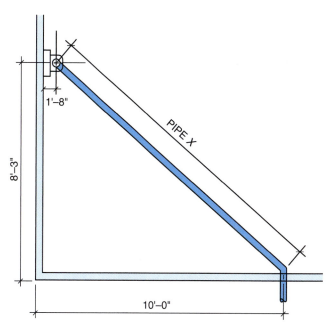

Figure 11–6

6-8-10 Method

Many things in construction are laid out with square corners, and the 6-8-10 method is a good way to check the squareness of those corners. The 6-8-10 method is simply an application of the Pythagorean theorem. Suppose we have a triangle with sides of 6 units and 8 units. We can calculate its hypotenuse by using $a^2 + b^2 = c^2$ as explained at the beginning of this chapter:

$$6^2 = 36$$
$$8^2 = 64$$
$$36 + 64 = 100$$

Thus the hypotenuse is the square root of 100, or 10. This is a 6-8-10 triangle. It is practical to know this if, say, we have two lines that we want to meet in a square corner. We can measure 6 feet along one line and 8 feet along the other line. When those two points are exactly 10 feet apart, the corner is square.

Problems

Supply the missing dimension in each of the following:

11.

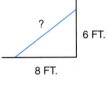

? 6 FT. 8 FT.

12.

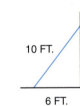

10 FT. ? 6 FT.

13.

10 FT. ? 8 FT.

14.

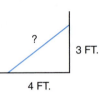

? 3 FT. 4 FT.

15.

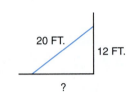

20 FT. 12 FT. ?

Chapter 12 | Combined Operations

OBJECTIVES

After completing this chapter, the student should be able to:

⊗ solve construction-related math problems involving combinations of the math topics covered in Chapters 5 through 11.

Construction problems do not always fall into neat categories like addition, fractions, linear measure, and so forth. When you work in construction, you are often confronted with situations in which you will have to use a variety of math skills and combine them in interesting new ways. That should not be a problem for you if you are able to do all the problems in the earlier chapters of this book.

Basic Principles

This chapter is primarily a collection of practice problems that require more than one step. If you can do the math for the individual steps, you can solve problems requiring these combined operations. There are two keys to solving a combined-operations problem: First, analyze the problem to determine what information is given and what you must calculate. Second, write down the steps you will go through. Very often one of the steps will be to get everything into the same units.

EXAMPLE ① What is the remaining area of a piece of sheet metal 3 feet wide by 5 feet long if a triangle with a base of 2 feet and a height of 15 inches is cut out of it? See Figure 12–1.

1. What do we already know?

   Length of sheet metal
   Width of sheet metal
   Base of triangle
   Height of triangle

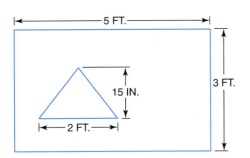

Figure 12–1

2. What are we looking for?

   Difference in area of the rectangle and the triangle

3. What are the steps we must do?

   Find the area of the rectangle. ($A = l \times w$)
   Find the area of the triangle ($A = 1/2\, b \times h$)

(Our given information includes feet and inches. We will need to either convert 15 inches to 1¼ feet or convert all foot dimensions of both parts to inches. Converting everything to inches makes for big numbers, but with a calculator, the math is easy.)

4. Convert the dimensions of the rectangle to inches.

   3 ft. × 5 ft. = 36 in. × 60 in.

5. Find the area of the rectangle.

   $A = l \times w$

   36 in. × 60 in. = 2,160 sq. in.

6. Convert the base of the triangle to inches.

   2 ft. = 24 in.

7. Find the area of the triangle.

   $A = 1/2\, b \times h$

   1/2 × 24 × 15 = 180 sq. in.

8. Subtract the area of the triangle from the area of the rectangle.

   2,160 sq. in. − 180 sq. in. = 1,980 sq. in.

9. Convert to square feet, because most of the dimensions were given in square feet.

   1 sq. ft. = 144 sq. in.

   1,980 ÷ 144 = 13.75, or 13¾ sq. ft.

Problems

Solve the following problems:

1. A fuel tank containing 240 gallons of fuel oil supplies an oil burner that consumes 1.3 gallons per hour. If the burner runs 1/3 of the time, how many hours will it take for the supply to be reduced to 100 gallons?

ESTIMATE

KOOL REFRIGERATION COMPANY
123 MAIN STREET
HOMETOWN, CA 12345

DATE: _____
SALES PERSON: _____
NO. 2404

EQUIPMENT & PARTS		LABOR			
DESCRIPTION/PART NO.	PRICE	DESCRIPTION	HOURS	RATE	AMOUNT
		EQUIP. & PARTS TOTAL			
		TAX			
		LABOR TOTAL			
		TOTAL ESTIMATE			

Figure 12–2

2. A customer asks for an estimate for the installation of a central air conditioner. A 3-ton unit is required. The estimate will show the estimated costs for parts and labor separately. Copy the form in Figure 12–2 to write your estimate. Include the following:
   a. 3-ton air conditioner $1.469.00
   b. Tubing, fittings, and hardware $62.50
   c. Labor (15 hours) $22.00 per hour
   d. Add 4 percent tax on equipment and hardware, but not on labor. Four percent would be 0.04 times the cost of the taxable portion of the estimate.

3. A crew of 3 can apply roofing at the rate of 260 square feet per hour. They are each paid $18 per hour. How much pay will each person receive for completing the roof shown in Figure 12–3?

4. How many cubic yards of concrete will be needed to build the foundation shown in Figure 12–4? All foundation walls are 8 feet 0 inches high. There is a

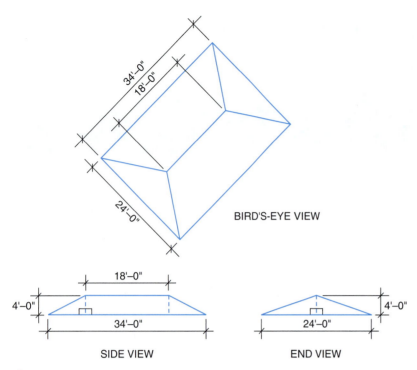

BIRD'S-EYE VIEW

SIDE VIEW

END VIEW

Figure 12–3

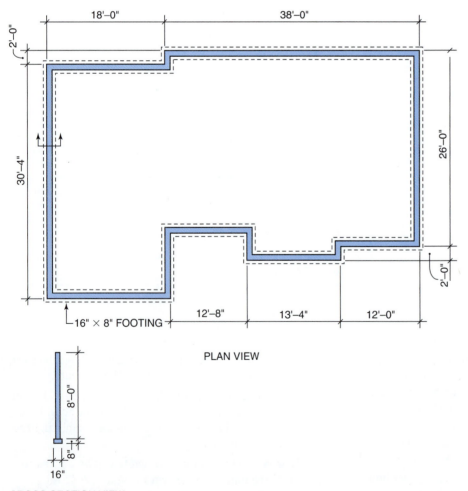

16" × 8" FOOTING

PLAN VIEW

CROSS-SECTION VIEW

Figure 12–4

16-inch-wide by 8-inch-thick concrete footing under the walls.

5. The diameter of a tank is 1.3 meters, and it is 3.1 meters long. How many gallons of fuel will the tank hold? One gallon is equal to 3,785 cubic centimeters. Round your answer to the nearest gallon.

6. The engine container in Figure 12–5 is filled with dry air at a pressure of 3 pounds per square inch. What is the total force against the inside of the top half of the container?

7. Find the length of part X for the special truss in Figure 12–6 to the nearest 1/8 inch.

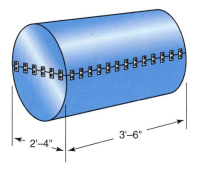

2'-4"   3'-6"

Figure 12–5

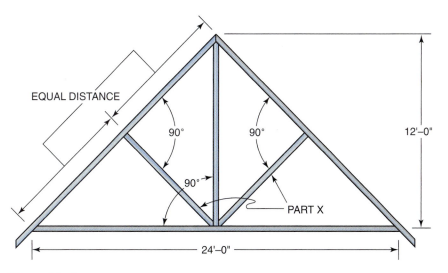

EQUAL DISTANCE

90°   90°

90°

PART X

12'-0"

24'-0"

Figure 12–6

Tools and Fasteners

Rex Keeler

TITLE

Field superintendent, Prairie Home Builders, Inc., in Lincoln, Nebraska

EDUCATION

Rex is a graduate of Southeast Community College, Milford Campus, in Milford, Nebraska. He has an associate's degree in building construction technology.

HISTORY

After graduating from SCC, Rex went to work for a general contractor in the commercial side of construction. Within five years Rex was promoted to job foreman and a few years later to superintendent. In 2001, Rex decided to get involved in the residential side of construction and is enjoying a successful career with a busy builder/developer.

ON THE JOB

As a superintendent, Rex is responsible for ensuring that all building disciplines meet the Unified Building Code, the contract documents are followed, and the schedule is maintained. Rex is also involved in a safety program within the company that helps to promote a safe and healthy work environment.

BEST ASPECTS

"The best aspects of construction for me are working outside—in and with the elements—and knowing that I've had my hand in creating buildings and structures that serve and help people and that will be there long after I am gone," says Rex.

CHALLENGES

Rex finds that the biggest challenges are *time* and *timing*. There never seems to be enough time. And time is money, so time management is a necessity, as well as a requirement for being successful.

Rex runs multiple projects in several developments throughout the city. Some projects need more of his time than others, so developing good time management skills and setting priorities are challenges that Rex strives to address daily.

IMPORTANCE OF EDUCATION

Rex believes that education is the first tool everyone should acquire. Education opens the windows and doors to the world. What could be more important than that!

FUTURE OPPORTUNITIES

"I believe that with knowledge and experience, an individual creates opportunities," says Rex. He is part of a good team that is dedicated to standards and principles that they all share.

WORDS OF ADVICE

"There is no such thing as a dumb question! If you don't know, ask. And never stop learning."

Chapter 13

Hand Tools— Selection, Use, and Care

OBJECTIVES

After completing this chapter, the student should be able to:

- read a ruler to within 1/16 inch.
- identify common tools and explain what they are used for.
- safely perform simple operations with common tools.

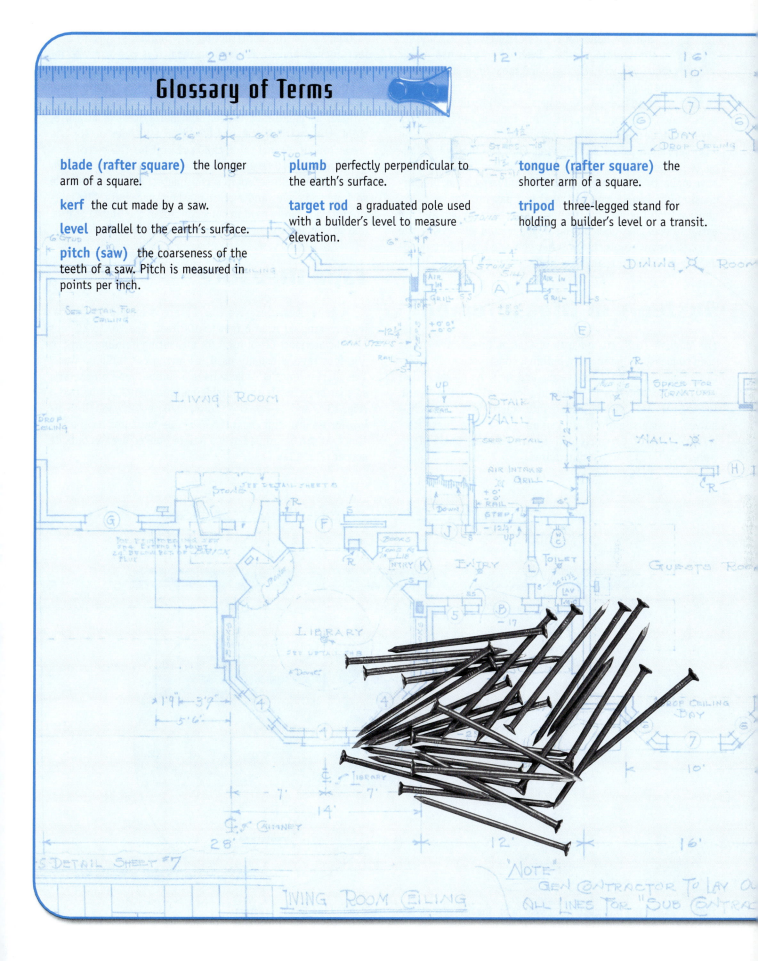

Glossary of Terms

blade (rafter square) the longer arm of a square.

kerf the cut made by a saw.

level parallel to the earth's surface.

pitch (saw) the coarseness of the teeth of a saw. Pitch is measured in points per inch.

plumb perfectly perpendicular to the earth's surface.

target rod a graduated pole used with a builder's level to measure elevation.

tongue (rafter square) the shorter arm of a square.

tripod three-legged stand for holding a builder's level or a transit.

ou will use hand tools for nearly every task you do in the building trades. Even when you use power tools, you will still use hand tools for many parts of the job. Knowing which tools to use for what purpose and how to use them properly will make your work easier, better quality, and safer. You will see hand tools advertised for unbelievably low prices, and you will see some with very high prices. By knowing what the tool will be expected to do, you will be able to make better decisions about how much to spend on a tool. Generally speaking, in purchasing hand tools, you get what you pay for. The more expensive tools are usually stronger, stay sharp longer, and are easier to use. That does not mean you should always buy the most expensive tools. Know what you need, and buy good-quality tools to do that job.

Systems of Measurement

The English (U.S. customary) system of linear measure is used to measure distances in building construction. This system is made up of inches, yards, and feet. As noted earlier in the book, there are 12 inches in a foot and 3 feet in a yard. Inches can be further divided into fractions of an inch—halves, quarters, eighths, and so on (Fig. 13–1). These fractional parts of an inch can be divided as finely as is necessary for the accuracy required.

Dimensions for construction are normally specified in feet, inches, and fractions of an inch. It is customary to reduce fractions to their simplest terms. For example, 26 and 6/16 inches is expressed as 2 feet 2⅜ inches. This is often written as 2 ft. 2⅜ in. or 2'-2⅜".

Also, as mentioned earlier, another system of measurement, the metric system, is based on multiples of 10. The meter (about 39 inches) is the base unit for the metric system. There are 10 decimeters or 100 centimeters in a meter. There are 10 millimeters in a centimeter. Tools are available with metric scales and markings where necessary. The metric system is rarely used in construction in the United States, and so it is not discussed further in this textbook.

Tape Measure

Steel tape measures (Fig. 13–2) are available in several lengths ranging from 6 feet to 100 feet. The 20- to 30-foot lengths are most often used for measuring building parts. A 100-foot tape is usually used for laying out building lines. The shorter lengths usually have a sliding hook on the end so that both inside and outside measurements can be taken (Fig. 13–3 and 13–4). The sliding hook moves enough to allow for its own thickness. Longer tape measures have a fitting that can be slipped over a nail or hooked over an outside corner (Fig. 13–5).

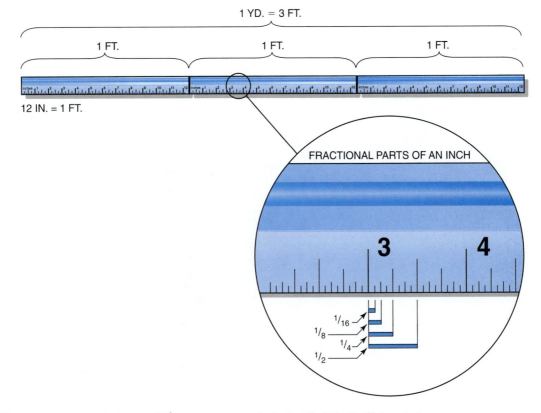

1 YD. = 3 FT.

1 FT.   1 FT.   1 FT.

12 IN. = 1 FT.

FRACTIONAL PARTS OF AN INCH

3   4

1/16
1/8
1/4
1/2

Figure 13–1 The U.S. customary system of linear measure, sometimes called the English system.

Figure 13–2 **Tape measures.** *Courtesy of Stanley Tools.*

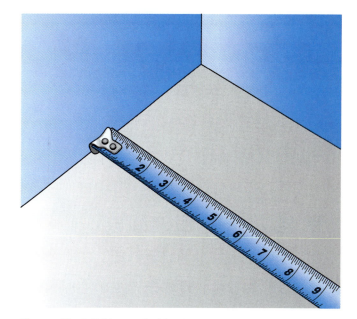

Figure 13–4 **Taking an inside measurement.**

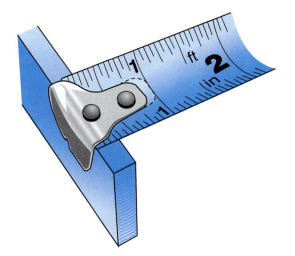

Figure 13–3 **The fitting on the end of a tape measure will slide to adjust for the thickness of the fitting.**

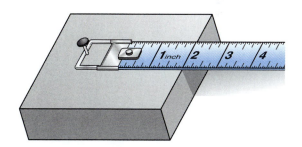

Figure 13–5 **The fitting on a long tape can slip over a nail or unfold to hook over a corner.**

Inspection and Defects

Tape measures are made of thin, fairly hard steel, and so it is quite possible to kink them. Rewind the tape carefully to ensure that it will not be kinked. Keep the tape clean and dry. Dirt and moisture will cause the marking on the face to wear away more quickly and will cause the rewind mechanism not to work properly. Inspect the fitting on the end of the tape to make sure it is not broken and it works properly.

Rafter Square

The rafter square, also called a framing square, is used for laying out or checking square corners (Fig. 13–6). The shorter part of the square is called the **tongue.** The longer part is the **blade.** Each edge of the rafter square is graduated with fractions of an inch, and the surfaces have tables that are useful for laying out rafters and braces.

Inspection and Defects

Rafter squares are durable and will last practically forever if handled with care, but dropping a rafter square or using it for anything other than what it is intended for can bend it. A bent square is useless.

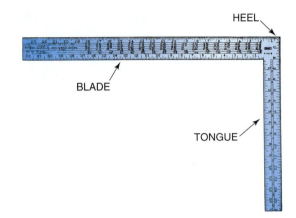

Figure 13–6 **Framing square.** *Courtesy of Stanley Tools.*

Figure 13–8 **Speed square.**

arm that can be set at any angle. Speed squares are convenient, because they are small, easy to carry, and quick to position and use.

Inspection and Defects

A speed square is a durable tool, but it cannot be bent, and it should not have nicks along the edge. Dropping the square or using it as a hammer or pry bar will surely bend and nick it.

Combination Square

The combination square (Fig. 13–9) has a movable head on a 12-inch blade. The head of the combination square has a right-angle surface, a 45° angle surface, and a small level. The combination square is handy in some applications, but not as tough nor as easy to carry as a speed square and not as large as a framing square.

Figure 13–7 **Stair gauges.** *Courtesy of Johnson Level & Tool Mfg. Co., Inc.*

Stair Gauges

Stair gauges are small fittings that can be clamped to the edge of the rafter square (Fig. 13–7). Stair gauges are used with a rafter square when several parts must be laid out with the same angles and dimensions.

Speed Square

A speed square (Fig. 13–8) is a triangular metal tool with several scales that are used for marking the cuts most often made on rafters. Some speed squares also have an adjustable

Figure 13–9 **Combination square.**

Inspection and Defects

The combination square has three separate parts: blade, head, and awl. All three parts should be in place, but even if the awl is missing, the rest of the square will work fine. Dropping the square or using it to pry can bend the blade, cause nicks in the blade that might prevent the head from fitting correctly, and damage the level.

Chalk Line Reel

A chalk line reel (Fig. 13–10) is used to mark long straight lines, such as to lay out a wall on a floor. The reel case, or chalk box, contains powdered chalk that coats the line as it is pulled from the reel. The chalk-covered line is then stretched tight while a point near its midpoint is pulled away from the surface and released (Fig. 13–11). When the line is

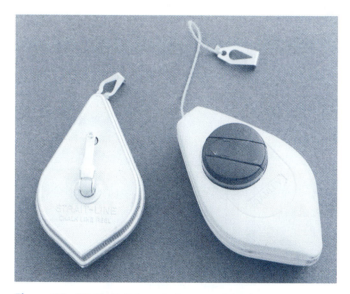

Figure 13–10 **Chalk line reel.** *Courtesy of Stanley Tools.*

Figure 13–11 **Snapping a chalk line.**

Figure 13–12 **Mason's line used to lay out building lines.**

released, it snaps against the surface to be marked, depositing a line of chalk on the surface.

Lines similar to that used in a chalk line reel are also used by masons to check the desired height of a course (row) of bricks or blocks. The line is stretched along the wall and checked for level. The bricks or blocks are leveled with this line as they are put in place. Mason's line is also used for laying out building lines (Fig. 13–12).

Inspection and Defects

Keep the chalk line clean and dry. Even a small amount of dirt can foul the line and cause it to miss spots when snapped. Any amount of water will contaminate the chalk. Inspect the chalk line reel before you use it to make sure the line is in good shape and the reel contains chalk.

6-8-10 Method and Checking Diagonals

Although these are not tools in the usual sense, they are valuable techniques for checking the squareness of large corners and rectangles. This technique was covered in Chapter 11, but it is so useful that it is repeated here. A triangle with sides of 6, 8, and 10 units of length contains a right angle (square corner). The units can be inches, feet, or any other unit of measure. This principle is used in building construction by measuring 6 feet along one side of a corner and 8 feet along the other side. If the corner is a perfect 90° angle, the distance between these points is exactly 10 feet (Fig. 13–13). You can check this with Pythagorean theorem, which you learned in Chapter 11.

The squareness of a square or rectangle can also be checked by measuring its diagonals. When all four corners are 90°, the two diagonals are equal in length (Fig. 13–14).

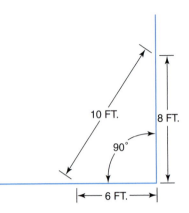

Figure 13–13 **Measure 6 feet along one side and 8 feet along the other side. Then check to see that these points are 10 feet apart.**

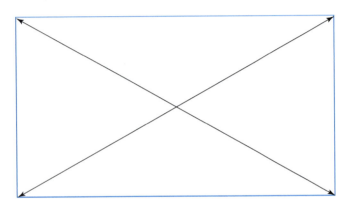

Figure 13–14 **If the diagonals are equal, all four corners are square.**

Spirit Level

A spirit level has one or more small transparent (usually acrylic) tubes, or vials, filled with mineral spirits for determining levelness or plumbness (Fig. 13–15). The word **level** actually refers to being parallel to the ground. To be **plumb** means to be perpendicular to the ground (Fig. 13–16). Spirit levels are available in a wide range of lengths, from the short torpedo level to an 8-foot level. The most common sizes are the torpedo level (about 9 inches), 2-foot level, and 4-foot level. There are also special types of spirit levels, such as the very small line level (Fig. 13–17).

CAUTION

CAUTION: The level is a delicate, precision tool. If it is dropped or jarred, the vials van be knocked out of alignment and the level will be useless.

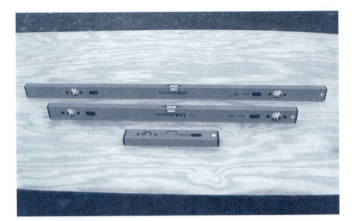

Figure 13–15 **A spirit level has one or more transparent vials containing a liquid and a bubble.**

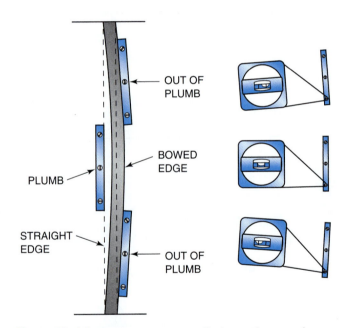

Figure 13–16 **Plumb means perpendicular to the ground.**

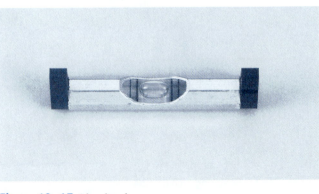

Figure 13–17 **Line level.**

Inspection and Defects

Look the level over before you use it, checking for visible signs of damage. If it looks like the level might have been dropped or otherwise abused, check it against a level that you know is in good condition to be sure that each vial is accurate.

Builder's Level

A builder's level is an instrument used in construction for measuring vertical distances over large horizontal areas, such as for the corners of a house foundation. The parts of the builder's level are shown in Figure 13–18. The functions of the parts are as follows:

- *Telescope* contains the lens, focusing adjustment, and crosshairs for sighting.
- *Telescope level* is a spirit level used for leveling the instrument when it is set up for use.
- *Clamp screw* locks the instrument in position so that it cannot be turned off the target.
- *Fine adjusting screw* makes fine adjustments to the left or right. (This cannot be seen in the figure.)
- *Leveling base* holds four leveling screws for leveling the instrument prior to use.
- *Protractor* is a scale graduated in degrees and minutes (a minute is 1/60 of a degree), used for measuring horizontal angles.

A builder's level is not a transit. A transit can be tilted to measure vertical angles (Fig. 13–19). A builder's level does not tilt.

Two accessories are required for most uses of a builder's level. The **tripod** is a three-legged stand that provides a stable, yet portable, base for the instrument. A **target rod** is a pole-like device with a scale graduated in feet and tenths of a foot. Each tenth of a foot is further divided into 10 parts, or 1/100ths, of a foot (Fig. 13–20). The builder's level is focused on the target rod to measure elevation.

Measuring Elevations

The following procedure can be used to determine the difference in elevation (height) of two ends of a drain pipe.

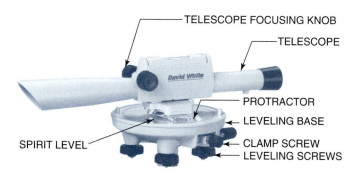

Figure 13–18 **Builder's level.** *Courtesy of David White.*

❶ Set the tripod up in a convenient place, where the level will have a clear line of sight over both ends of the pipe.

CAUTION

CAUTION: If the level will need to remain in place while others work in the area, make sure that it is not near where other work will be done.

CAUTION

CAUTION: Set the feet of the tripod solidly into the ground so that they will not move. If the tripod is being set up on a smooth surface, such as a plywood deck, fasten a small chain between the legs so that they will not spread further apart than you want them to.

❷ Set the instrument on top of the tripod, and hand-tighten the clamp screw.

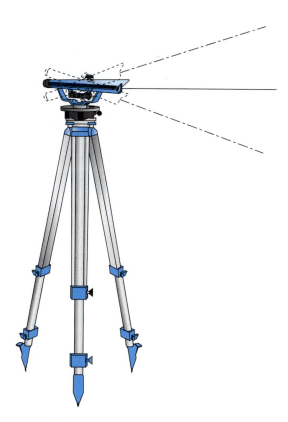

Figure 13–19 A transit is similar to a builder's level, but it can be tilted to measure vertical angles.

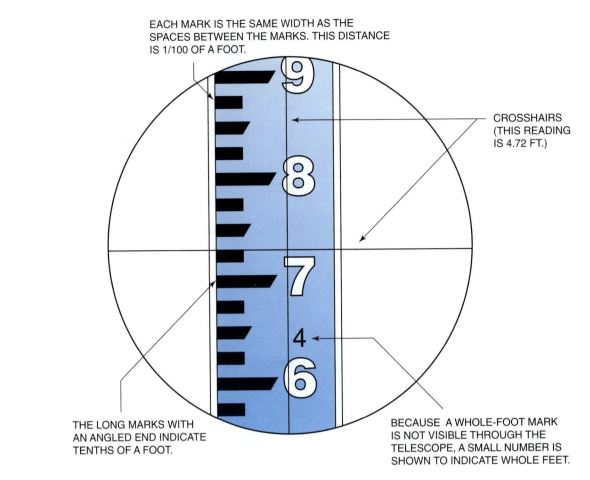

EACH MARK IS THE SAME WIDTH AS THE SPACES BETWEEN THE MARKS. THIS DISTANCE IS 1/100 OF A FOOT.

CROSSHAIRS (THIS READING IS 4.72 FT.)

THE LONG MARKS WITH AN ANGLED END INDICATE TENTHS OF A FOOT.

BECAUSE A WHOLE-FOOT MARK IS NOT VISIBLE THROUGH THE TELESCOPE, A SMALL NUMBER IS SHOWN TO INDICATE WHOLE FEET.

Figure 13–20 Graduations on a target rod.

CAUTION

CAUTION: The level is a delicate instrument. Keep it in its case when it is not being used, and handle it carefully.

③ Turn the leveling screws down so that they contact the tripod plate.

④ Turn the telescope so that it is over one pair of leveling screws. Adjust these two screws to make the telescope level. (Hint: If you turn both screws at once, the level will always move in the same direction as your left thumb.)

⑤ Turn the telescope so that it is over the other pair of leveling screws. Adjust these two screws to make the instrument level.

⑥ Repeat steps 4 and 5, alternating over both pairs of leveling screws until the instrument is level in both positions. Be careful not to touch the tripod with your foot. The slightest jar of the tripod will cause the instrument to be knocked out of level.

⑦ Have a partner hold the target rod over one end of the pipe while you sight through the telescope and focus the crosshairs on the target rod. Write down the measurement seen through the telescope.

⑧ Have your partner move the target rod to the other end of the pipe. Carefully rotate the instrument, without touching the tripod, so that you can focus on the target rod over this end of the pipe. Write down the reading at this end.

⑨ Subtract the smaller reading from the larger one to find the difference in elevation from one end of the pipe to the other.

Laser Level

A *laser* is a device that gives off a very focused beam of light—so focused that it can be seen as a small dot over great distances. A laser level is a leveling instrument that uses laser light instead of eyesight focused through a telescope. There are many manufacturers and types of laser levels. The more expensive laser levels are usually self-leveling, but the less expensive ones, which are very common in construction, must be leveled on a tripod much the same as a builder's level is leveled (Fig. 13–21).

Figure 13–21 Laser level.

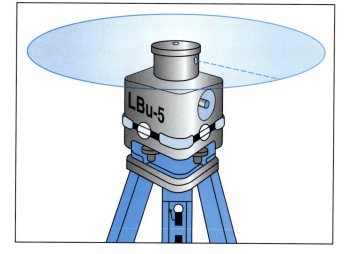

Figure 13–22 **The laser rotates in a complete circle, creating a level beam of light.** *Courtesy of Laser Alignment, Inc.*

Measuring Elevations

Once the instrument is leveled, the laser is turned on. It gives off a narrow beam of light. The laser level rotates this beam of light through a complete circle (Fig. 13–22). As it rotates, it creates a red line of light that strikes everything in its path at the same elevation.

The laser beam is difficult to see outdoors in bright sunlight. To detect the beam, a battery-powered sensor, called a receiver or detector, is attached to the target rod. Most sensors have a visual display, with a selectable audio signal to announce when it is close to or right on the laser beam (Fig. 13–23). Specially designed laser targets are available for some jobs, such as installing suspended ceiling grids.

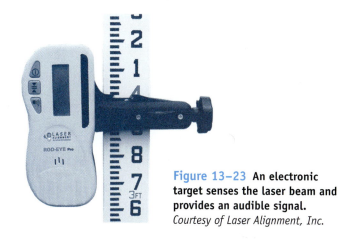

Figure 13–23 **An electronic target senses the laser beam and provides an audible signal.** *Courtesy of Laser Alignment, Inc.*

CAUTION

CAUTION: All laser instruments are required to have warning labels attached (Fig. 13–24). Only trained workers should set up and use laser instruments, and the following safety rules are to be followed:

- **Never look directly into a laser beam.**
- **Never view a laser beam with an optical instrument (builder's level).**
- **When possible, set up the laser level so that it is above or below eye level.**
- **Turn the laser off when not in use.**
- **Do not point a laser beam at another person.**

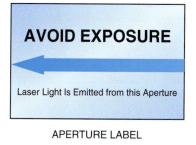

AVOID EXPOSURE

Laser Light Is Emitted from this Aperture

APERTURE LABEL

DANGER

LASER LIGHT - AVOID
DIRECT EYE EXPOSURE.

HELIUM NEON LASER
5 MW MAXIMUM OUTPUT
CLASS III a LASER PRODUCT

WARNING LABEL

Figure 13–24 **All lasers are required to have warning labels.** *Courtesy of Laser Alignment, Inc.*

Plumb Bob

A plumb bob (Fig. 13–25) is a pointed weight that can be attached to a string. Plumb bobs are made of brass or steel. Brass plumb bobs usually have a steel tip, because the softer

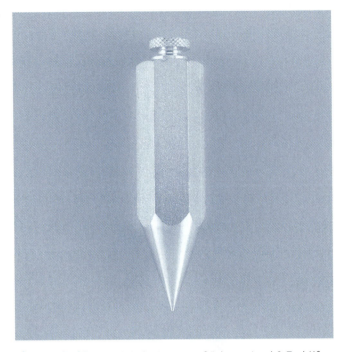

Figure 13–25 **Plumb bob.** *Courtesy of Johnson Level & Tool Mfg. Co., Inc.*

brass could be damaged easily. They are available in weights from 5 ounces to 24 ounces. Gravity causes the plumb bob to hang perfectly vertical or perpendicular to the earth's surface. A perfectly vertical line is said to be plumb. A plumb bob can be used to show where the bottom of a post should go to support a beam or to line a tripod up over a precise point.

Hammers

Claw Hammer

The familiar hammer that a carpenter uses is a claw hammer, named after the claw on the opposite side from the face of the hammer (Fig. 13–26). There are two styles of claw hammers, curved claw and straight claw. Claw hammers are also available in different weights, with 16 ounce and 20 ounce being the most common. Framing carpenters usually use 20-ounce, straight-claw hammers. The extra weight helps drive the large nails used in framing, and the straight claw is most useful for prying out nails or separating nailed boards. Trim carpenters often prefer a 16-ounce hammer for the smaller finishing nails they use and because the lighter hammer does less damage to the surrounding wood. Workers in other trades use whichever hammer weight best suits the type of work they do most.

To use a claw hammer to drive nails, grip it near the end of the handle with one hand. Hold the hammer firmly, but do not squeeze it. Use your whole arm to swing the hammer in a complete arc, keeping your eyes focused on your target—the head of the nail. Avoid the temptation to

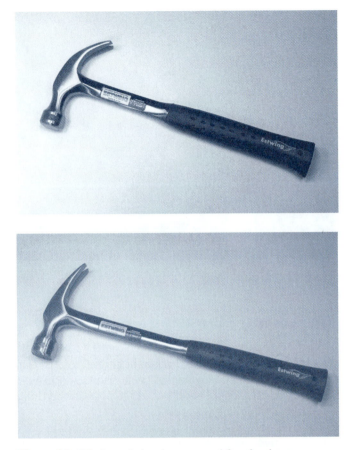

Figure 13–26 **Curved-claw hammer and framing hammer.**

use two hands. The extra hand will cause you to miss the nail head.

CAUTION

CAUTION: Wear safety glasses when hammering. Nail heads can break off, and a nail that is not hit squarely can fly out from under the hammer.

Inspection and Defects

Inspect the hammer for damage to the handle or the head before you use it. If the handle is wood, look for signs of cracks or splintered wood. Also check to see that the head is firm on the handle. If the handle is fiberglass, look for splintered fiberglass. Check the head over, especially looking for chips in the hardened face of the hammerhead. Do not use a hammer with any of these defects.

Bricklayer's Hammer

The bricklayer's hammer, also called a mason's hammer or simply brick hammer, has a square face and a chisel-like cutting edge (Fig. 13–27). Like claw hammers, bricklayer's

Figure 13–27 **Bricklayer's hammer.** *Courtesy of The Stanley Works.*

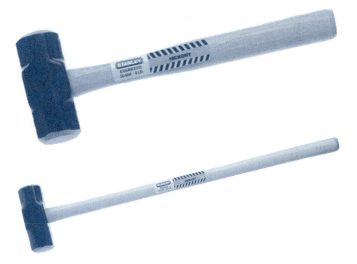

Figure 13–28 **Sledgehammers.** *Courtesy of The Stanley Works.*

hammers are available with wood, steel, or fiberglass handles and in different weights. The flat face of the hammer is used for occasional nailing and for striking cold chisels. The other end of the hammerhead can be used to cut bricks and concrete blocks. The brick or block is scored on all four sides by striking with the cutting edge, and then it is broken with a final blow on the score line.

CAUTION

CAUTION: Always wear safety glasses when using a bricklayer's hammer.

Inspection and Defects

Inspect a bricklayer's hammer for the same defects as for a claw hammer.

Sledgehammer

Sledgehammers are used for driving stakes and breaking up hard materials. Sledgehammer heads are available from 2 pounds to 20 pounds. Sledgehammers are also available with short handles for one-handed use or longer handles for use with two hands (Fig. 13–28).

CAUTION

CAUTION: Sledgehammers are capable doing serious injury. Obey the following safety rules when using any sledgehammer:

- **Check to see that the head is secure on the handle before use. Do not use any hammer with a loose head.**
- **Do not use a sledgehammer with splits or other damage to the handle.**

- **Do not attempt to hold an object with your hands while it is being struck with a sledgehammer. Start stakes with a claw hammer or mallet first.**
- **Wear safety glasses, steel-toe boots, and gloves while using a sledgehammer.**

Inspection and Defects

Inspect a sledgehammer the same as you would a claw hammer—both the handle and the head. Do not use a damaged sledgehammer.

Bars and Nail Pullers

Several types of pry bars are used in construction (Fig. 13–29). Wrecking bars, flat bars, and cats paw's all have notches in the ends that can be used for pulling nails. A bar can apply much greater force to remove a nail than can a claw hammer. The straighter end is sometimes used to pry pieces apart. It is also used in other places where leverage is needed.

Inspection and Defects

Bars and nail pullers are subject to a lot of rough use. If they have been struck with a hammer, they might have mushroomed metal at that point. Any sharp edges should be removed with a grinder before a bar is used.

Screwdrivers

Screwdrivers can have slotted, Phillips, torx, or square tips (Fig. 13–30). There are many other types of screwdriver tips, but these are the most common. At one time, slotted screws

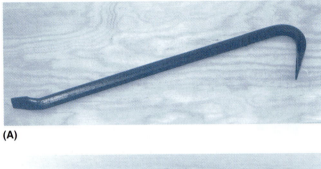

(A)

(B)

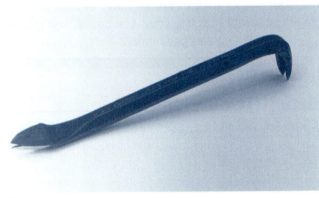

(C)

Figure 13–29 **(A) Wrecking bar, (B) flat bar, and (C) cat's paw.**

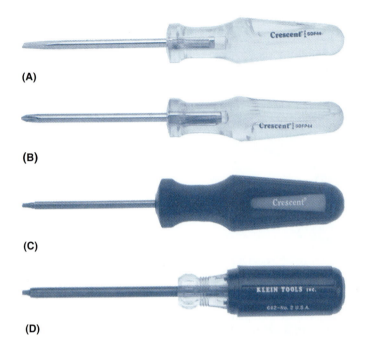

(A)

(B)

(C)

(D)

Figure 13–30 **Screwdrivers: (A) slotted, (B) Phillips, (C) torx,** *Courtesy of COOPER Tools,* **(D) square or Robertson,** *Courtesy of Klein Tools, Inc.*

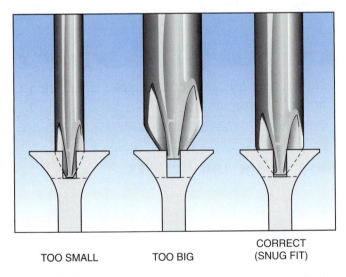

TOO SMALL   TOO BIG   CORRECT
(SNUG FIT)

Figure 13–31 **The screwdriver must be the right size to fit the fastener.**

were the standard, with only occasional use of the Phillips screw. Today, Phillips and square-drive screws are the most common in industry. The torx design is not used in construction as much as in the automotive and machine industries. All types of screwdrivers are available in different sizes. The most common mistake people make in using screwdrivers is to use the wrong size (Fig. 13–31). A screwdriver that fits the screw head properly is less apt to slip out and mar the surrounding wood or to damage the head of the screw. The second most common mistake is using a screwdriver with a damaged tip.

Inspection and Defects

Inspect screwdrivers for the condition of the tip. A slotted screwdriver should have a well-shaped end, with square corners, and the tip should be straight. Phillips, square, and other internal screwdrivers should have well-shaped ends, with no noticeable nicks, gouges, or rounded edges. The handle should be in good shape. Some screwdrivers have been struck with hammers, turned with pliers, or used as a pry bar. These uses almost always damage the screwdriver.

Pliers

There are many types of pliers. The ones that are most often used in construction are shown in Figure 13–32. Each has its own advantages in certain situation. Needle-nose pliers are

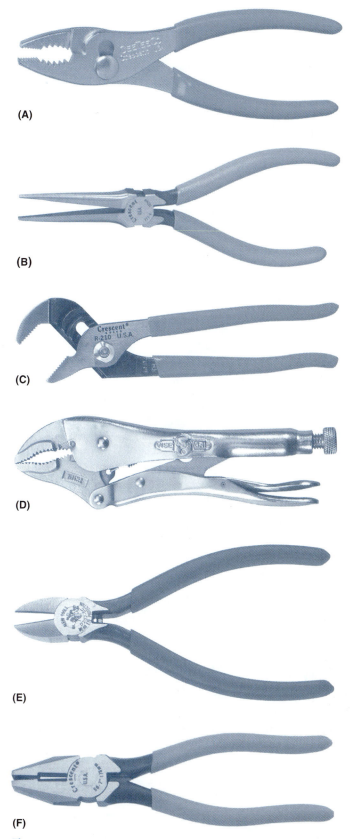

useful for handling small, delicate objects, while vise-grip pliers apply maximum gripping strength. Electrician's pliers have a cutting section, for cutting wires, and a gripping section; but side-cutting pliers are used only for cutting. Slip-joint or channel-lock pliers can be adjusted for the widest range of sizes. All types of pliers are available in several sizes. Use the type and size pliers that are right for the job.

Inspection and Defects

Pliers are usually pretty tough. Unless they have been badly misused, they will generally work as intended for gripping. Do not expose pliers to high temperature; for example, do not hold a piece in the flame of a torch. Do not hammer on pliers or use them as a hammer. Electrician's pliers and other cutting pliers may have damaged or dull cutting edges. They are not intended for cutting hardened wire, and doing so will damage the cutting edges. You might not notice this until you try to use them. They cannot practically be sharpened, and so the best thing to do with dull cutting pliers is to replace them.

Wrenches

Socket Wrenches

A socket wrench consists of a socket and a handle, Figure 13-33. Sockets are available in sizes from 1/4 inch to 2 inches and from 3 mm to 50 mm. The most common

Figure 13–32 Pliers: **(A) common slip-joint, (B) needle nose, (C) channel-lock,** *Courtesy of COOPER Tools,* **(D) vise-grip,** *Courtesy of Irwin Industrial Tool Company,* **(E) side-cutting,** *Courtesy of Klein Tools, Inc.,* **(F) electrician's,** *Courtesy of COOPER Tools.*

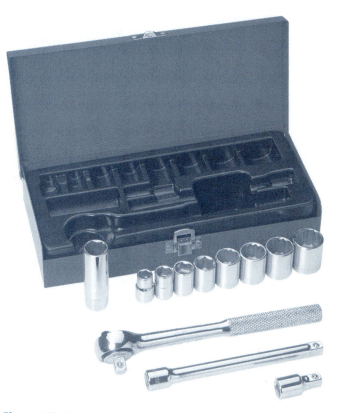

Figure 13–33 **Socket wrench set.** *Courtesy of Klein Tools, Inc.*

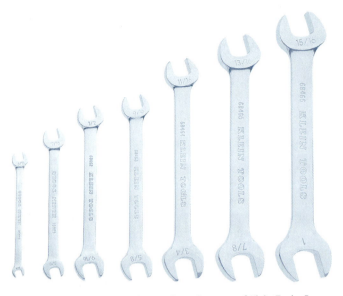

Figure 13–34 **Open-end wrenches.** *Courtesy of Klein Tools, Inc.*

sizes for construction work are 3/8 inch to 7/8 inch. The most common socket handle types are flex handles and ratchet handles. The flex handle is used where maximum force must be applied. The ratchet handle allows the user to move the handle back and forth without having to take the socket off the nut and reposition it. Socket handles and the sockets themselves are made for 1/4-inch drive, 3/8-inch drive, 1/2-inch drive, and 3/4-inch drive. The most popular sizes are 3/8-inch and 1/2-inch drive.

Open-End Wrenches

Open-end wrenches (Fig. 13–34) usually have different sizes at each end. Some open-end wrenches are the same size on both ends, but have the ends set at different angles. This can be very useful for hard-to-reach nuts and bolts.

CAUTION

CAUTION: Take the time to find the right-size wrench. A wrench that is just slightly too big can slip and cause damage to the nut or bolt and can cause painful injury to your knuckles. Many jobs have taken far longer than they should because someone tried to make do with a loose-fitting wrench and damaged the head of a bolt.

Box-End Wrenches

Box-end wrenches (Fig. 13–35) have a different size at each end of the wrench. The end forms a complete circle and has either 6 or 12 points. Twelve-point wrenches are generally the more versatile of the two. Box wrenches are less apt

to slip on the bolt head or nut than are open-end wrenches, but they cannot always be used.

Figure 13–35 **Box-end wrench.**

Nut Drivers

Nut drivers look like a cross between a screwdriver and a socket wrench (Fig. 13–36). They are available in sizes to fit most smaller-size nuts and bolts—both U.S. customary and metric. It is very difficult to apply large amounts of twisting force, called torque, to a nut driver, and so nut drivers are not generally made to fit larger sizes.

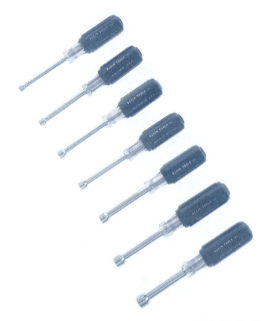

Figure 13–36 **Nut drivers.** *Courtesy of Klein Tools, Inc.*

Adjustable Wrenches

An adjustable wrench has a fixed jaw and a movable jaw (Fig. 13–37). The opening of the wrench is adjusted by turning the adjusting screw, which moves the movable jaw. To use an adjustable wrench, apply the force toward the movable

Figure 13–37 **Adjustable wrench.** *Courtesy of COOPER Tools.*

jaw. Adjustable wrenches come in sizes from 4 inches to 2 feet. A 10- or 12-inch adjustable wrench is a popular size.

Pipe Wrenches

A pipe wrench (Fig. 13–38) is another type of adjustable wrench. The jaws of a pipe wrench have teeth so that they can grip the round surfaces of pipes. As with an ordinary adjustable wrench, the force should be applied toward the movable jaw.

Figure 13–38 **Pipe wrench.** *Courtesy of Klein Tools, Inc.*

Basin Wrenches

A basin wrench (Fig. 13–39) is used to reach the large retaining nuts that hold faucets in place on sinks. The basin wrench is a self-adjusting wrench on a long tee handle. The wrench end pivots so that it can be flipped over and used to turn the nut in the opposite direction. Some basin wrenches have a telescoping handle that can be extended for longer reach or shortened for easier use.

Inspection and Defects

Although their designs are different, most wrenches are susceptible to the same damage, usually from one of the following:

- Having been used as a hammer. Using a wrench as a hammer can cause the metal to be deformed and might damage the gripping surfaces or the handle. A hammer cannot be all that far away!
- Having been hammered on themselves. Hammering on a wrench puts sudden impact where the wrench was not meant to absorb high stresses. It can crack or break the wrench, leave sharp burred edges, and stretch the jaws of the wrench so that it will not fit any standard size.
- Using the wrong-size wrench. A wrench that is too small simply will not fit on the bolt head or whatever part is to be turned. A wrench that is slightly too big might appear to work, but it can slip. When it does, it can cause painful scrapes on your knuckles; it may round

the corners on the bolt head so that no wrench will work well; and it may damage the gripping surfaces of the wrench.

- Being forced beyond their working limit. If a wrench is forced way beyond its working strength, it might crack. Do not use a piece of pipe to extend the length of a wrench for more leverage.

Hacksaw

A hacksaw consists of a metal frame to hold a hacksaw blade and a handle (Fig. 13–40). Hacksaw frames can be adjusted to hold different-length blades. The blade is installed with the teeth pointing away from the handle, and then the adjustment is tightened to make the blade taut. Since the teeth point away from the handle, the cutting is done on the forward stroke of the saw. The saw should be lifted slightly off the work on the return stroke.

Inspection and Defects

Inspect the hacksaw blade to make sure no teeth are missing and the teeth are sharp. Dull hacksaw teeth will not have uniform sharp points. Inspect the frame of the hacksaw to make sure it holds the blade properly and the blade is tight, with the teeth pointing away from the handle.

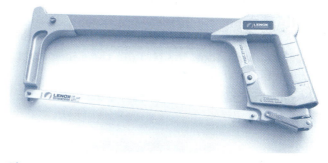

Figure 13–40 **Hacksaw.**

Handsaw

Handsaws are used for sawing wood. Different arrangements of teeth are needed for different types of sawing. The coarseness of a saw is called its **pitch.** Saw pitch is measured in points per inch (Fig. 13–41). This is the number of

Figure 13–39 **Basin wrench.** *Courtesy of Rigid Tools.*

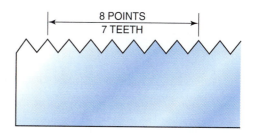

Figure 13–41 Saw pitch is measured in points per inch.

points of saw teeth in 1 inch of the blade. A hacksaw might have 32 teeth per inch, but a wood saw will have fewer points per inch.

The cut made by the saw is called a **kerf.** It is often very important to remember that a saw kerf has width. A kerf made by a typical wood saw is usually about 1/8 inch wide.

Crosscut Saws

Crosscut saws are used to saw wood across its grain. Crosscut saws for general purposes usually have 8 to 12 points per inch. The teeth are filed (sharpened) at a slight angle so that they come to sharp points at alternate sides. The top third of each tooth is set (bent slightly) to the sharp-pointed side (Fig. 13–42). Set is needed in most saws, so the kerf will be slightly wider than the thickness of the saw blade. The kerf is the actual cut made by the saw. The kerf is as wide as the width of the set in the saw blade. Having a kerf that is wider than the thickness of the body of the saw blade prevents the saw from binding in the kerf.

As the crosscut saw is pushed forward, the first parts to contact the wood are the sharp points at the sides of the kerf. This cuts the wood fibers off before they are removed. As the teeth cut deeper, the body of the tooth removes the sawdust from the kerf.

CAUTION

CAUTION: A properly sharpened saw should be handled with care. Keep your hands away from the saw teeth and never allow the teeth to come in contact with metal or masonry.

Ripsaws

A ripsaw is used to cut with the grain of the wood. This is called ripping. Ripsaws generally have 4 to 8 points per inch. Ripsaw teeth are filed straight across the blade, and so each tooth is shaped like a little chisel (Fig. 13–43). To prevent binding, ripsaw teeth are set much like crosscut saw teeth.

As the ripsaw teeth contact the wood, the full width of each tooth chisels out a small amount of wood. Because the sawing is in the same direction as the fibers of the wood, it is not necessary to cut the fibers off before removing them. Ripsaw teeth are usually larger than crosscut saw teeth, because they must remove large amounts of complete wood fibers instead of short, cutoff pieces.

Inspection and Defects

Visually inspect the handsaw to see that the teeth appear to be sharp (uniform, sharp points on every tooth) and that the handle has no cracks and is firmly attached to the blade. If the saw has not been stored properly, it might be dull from having contacted hard objects.

Using a Handsaw

1. Place the wood to be sawed on a stable rest, such as two sawhorses, so that the saw will not come in contact with anything else. In most cases you will be able to hold the wood securely in place with your left hand, but the wood can be clamped in place if necessary.
2. Use a square or other straightedge to make a pencil mark where you want the cut to be made.
3. Hold the saw in your right hand if you are right-handed (reverse if you are left-handed) in a loose but controlled grip. Do not squeeze the saw.

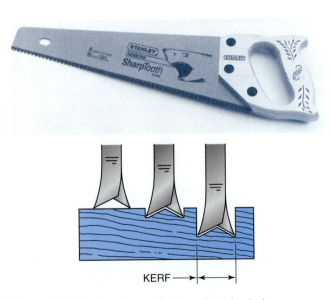

Figure 13–42 **Crosscut saw.** *Courtesy of Stanley Tools.*

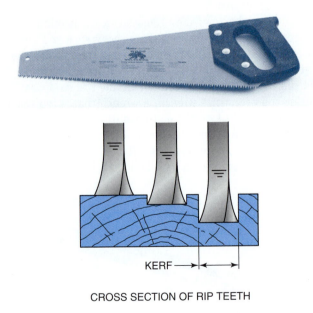

CROSS SECTION OF RIP TEETH

Figure 13–43 **Ripsaw.** *Courtesy of Stanley Tools.*

4. Position your body where you will be comfortable and out of the way of the saw. Place the saw on the edge or end of the board, beside the pencil line, but on the scrap side of the line. Remember, your kerf will be about 1/8 inch wide.

5. Hold the saw at about a 45° angle with the piece of wood, and push the saw forward, allowing the weight of the saw to push the teeth against the wood. The cutting action will be on the forward stroke.

6. Apply very slight twisting pressure as necessary to keep the kerf on the scrap side of the line.

7. When you near the end of the cut, make sure the scrap is supported so that it will not tear away from the piece and leave splinters as it falls.

Coping Saw

Sometimes in fitting molding to inside corners, carpenters cut the profile of one piece of molding on the end of the other piece. This is called coping. To make intricate cuts, a coping saw is used. A coping saw has a small frame that holds a thin, fine-pitch crosscut blade (Fig. 13–44). Coping saws are useful for all kinds of sawing where intricate shapes are involved. The coping saw is one of the few handsaws in which the blade is installed with the teeth pointing toward the handle. The cutting action occurs as the saw is pulled toward you.

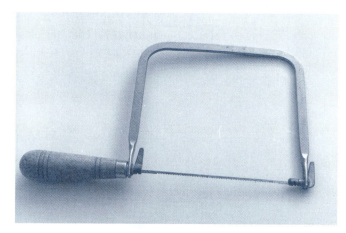

Figure 13–44 Coping saw.

Wallboard Saw

The wallboard saw (Fig. 13–45) is used for cutting small openings in wallboard. Wallboard saws have fairly coarse teeth, and so they cut gypsum wallboard quickly. They are narrow enough to make sharp curves.

CAUTION

CAUTION: When cutting holes in wallboard, make sure you know what is behind the wallboard. If the wallboard is installed on a wall, lock out and tag out the circuits that are buried in that wall. One stroke of a wallboard saw can cut an electric cable or puncture a copper pipe.

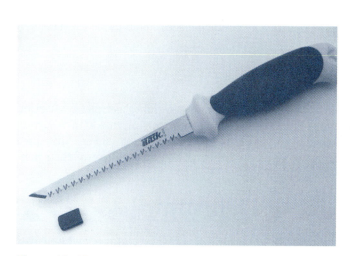

Figure 13–45 Wallboard saw.

Utility Knife

A utility knife (Fig. 13–46) is useful for many tasks, including scoring gypsum wallboard, cutting the backing on insulation, cutting shingles, and so on. There are many styles of utility knives, but most have a similar arrangement for storing extra blades in the handle.

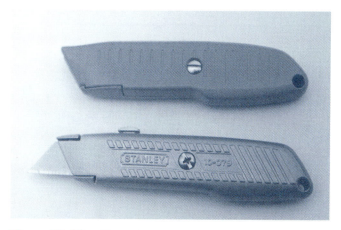

Figure 13–46 Utility knife.

CAUTION: Do not try to use a utility knife with a dull blade. It will not make a clean cut, it is difficult to make a straight cut with a dull knife, and it is dangerous. Do not cut toward yourself.

Inspection and Defects

Check to see that the utility knife has a clean, sharp blade. If not, there may be spare blades stored inside the handle.

Snips

Aviation snips are used to cut thin sheet metal. Aviation snips are designed for straight cuts or cuts with relatively gradual curves, left curves, and right curves. The three types of snips can be identified by the color of their handles (Fig. 13–47).

CAUTION: Wear gloves and safety glasses when cutting sheet metal. The cut edges can be very sharp.

Inspection and Defects

Snips should only be used to cut thin sheet metal. If you have to force the snips to make them cut, you may be trying to cut too thick a piece of sheet metal. Forcing the snips into heavy metal can spring the blades, causing them not to mate properly. Using snips to cut wire can nick the blades, causing them to leave a ragged edge. It is not practical to repair damaged snips.

Pipe and Tubing Cutters

The tubing cutter shown in Figure 13–48 is used to cut aluminum, copper, brass, and steel tubing and plastic pipe. It has two rollers in the lower jaw and a round cutter in the upper jaw. The cutter can be advanced toward or retracted away from the rollers by turning the knob. The cutting wheel that comes with most tubing cutters is designed to cut copper, brass, and aluminum tubing. To cut steel or plastic, the cutting wheel should be changed to one made for that material. A pipe cutter is a larger version of the same tool.

To use a tubing or pipe cutter, mark the tubing where it is to be cut. Turn the knob counterclockwise enough to allow the cutting wheel and rollers to fit over the pipe. Turn the knob clockwise just until the cutting wheel contacts the pipe or tubing at the cut-off mark. Tighten the knob about 1/16 of a turn beyond the point where it makes contact. You should feel a firm pressure, but the knob should not be as tight as you can make it. Rotate the cutter around the pipe or tubing. At the end of each revolution, tighten the knob another 1/16 of a turn. Continue this until the tubing or pipe is completely cut.

The triangular piece of metal on the back of the tubing cutter is a reamer. It is used to ream, or smooth, the inside edge of the cut end of the tubing. Position the point of the triangle in the end of the tubing. With moderate pressure against the end of the tubing, turn the reamer so that its edges cut away any burrs that are left in the cut end.

The type of tubing cutter shown in Figure 13–49 is made for cutting only plastic pipe. It is actually a shear with jaws to support the plastic pipe so that it will cut cleanly. Do not use a plastic pipe cutter to cut anything other than the plastic pipe for which it is intended. To do so will damage the cutting edges.

Figure 13–47 Right-, straight-, and left-cutting aviation snips.

Figure 13–48 Tubing cutter. *Courtesy of Rigid Tools.*

Figure 13–49 **Plastic pipe cutter.** *Courtesy of Rigid Tools.*

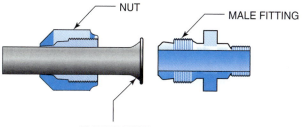

Figure 13–50 **Flared tube and fitting.**

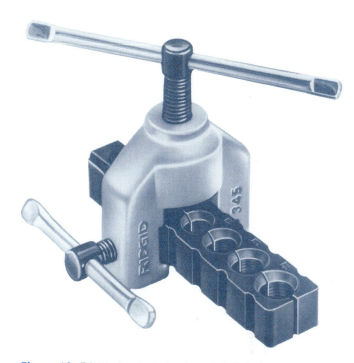

Figure 13–51 **Flaring tool.** *Courtesy of Rigid Tools.*

Inspection and Defects

Pipe and tubing cutters are actually quite rugged, but their cutting edges can be dulled or damaged if they are used to cut anything other than the materials for which they are intended. Inspect the cutting wheel or shear edges for nicks or signs of apparent damage. If they are damaged, the tool will be difficult to use and will not produce a clean cut. The cutting edges or cutting wheels in most tubing cutters are replaceable. Replace them if they are damaged. Also check to see that both rollers roll smoothly and that the cutting wheel is not loose on its axle.

Flaring Tool

One popular type of fitting for metal tubing is the flare fitting (Fig. 13–50). To use this type of fitting requires flaring the end of the tubing with a flaring tool like the one shown in Figure 13–51. The flaring tool consists of three parts. The first part is a pair of flaring bars that, when clamped together, create holes the right size for various sizes of tubing. The top of each of these holes has a small chamfer (the top of the hole is flared out wider than the main body of the hole). The other two parts of the flaring tool are the yoke, which clamps around the flaring bars, and the feed screw. The bottom of the feed screw is fitted with a cone-shaped end that matches the angle of the chamfers in the flaring bars (45°).

Operating Instructions

1. Cut and ream the tubing.
2. Back off the feed screw and loosen the yoke to permit the flaring bars to slide freely through the yoke.
3. Insert the tubing into the proper-size opening and close the flaring bars.
4. Push the tubing up from the bottom until it is just even with the top of the flaring bars.
5. Slide the yoke into place over the tube. (Most flaring tools have a mark on the yoke and a corresponding mark on the flaring bars to indicate when the yoke is

properly lined up.) Tighten the yoke in place, locking the flaring bars around the tube.
6. Tighten the feed screw until the tubing is flared against the sides of the flaring bar. A drop of oil on the fitting on the end of the feed screw may help produce a smoother flare.
7. Back off the feed screw as far as it will go. Loosen the yoke and slide it back so that the flaring bars can be opened.

CAUTION

CAUTION: Be careful to ensure that the angle of the flare matches the angle on the fitting. In plumbing and HVAC, 45° flares are standard. In automotive and machinery, 37° flares are standard. The two cannot be mixed.

Swaging Tools

Soft-metal tubing like copper and aluminum can be joined by expanding the end of one piece enough so that the other piece will fit inside and can be soldered without fittings. This is called swaging. The simplest type of swaging tool is a punch-like tool that can be forced into the end of the tubing, thereby expanding it to the size of the tool. A set of swaging tools for common sizes of tubing is shown in Figure 13–52. These swaging tools are driven into the end of the tubing with a hammer. Hard-steel tubing and plastic pipe cannot be easily swaged. Do not use swaging tools as punches or drifts. The smooth surface of the swaging tool must be maintained so that it will produce a good-fitting joint that can be soldered easily.

Figure 13–52 **Swaging tools.** *Courtesy of Rigid Tools.*

Review Questions

Select the most appropriate answer where applicable.

Match the uses in column II with the tools in column I.

Column I

1. 16-ounce, curved-claw hammer
2. Cat's paw
3. Plumb bob
4. Chalk line
5. 2-foot level
6. Builder's level
7. Square-head screwdriver
8. Electrician's pliers
9. 10-inch adjustable wrench
10. Nut driver
11. Crosscut saw
12. Coping saw
13. 20-ounce, straight-claw hammer
14. Rafter square
15. 18-inch pipe wrench
16. Utility knife

Column II

a. Sawing with the grain of wood
b. Sawing sharp curves in wood
c. Scoring gypsum wallboard
d. Checking square corners
e. Tightening and loosening small nuts and bolts
f. Checking and measuring elevations over long distances
g. Used by framing carpenters for driving large nails
h. For cutting wire and gripping parts
i. Used by finish carpenters for driving nails
j. Marking a straight line for many feet
k. Checking to make sure a line is perfectly vertical
l. Turning a screw-on fitting on a pipe
m. Pulling nails
n. Checking a window header to make sure it is level
o. Sawing across a board
p. Tightening large bolts

17. List three safety rules for using laser levels.
   a.
   b.
   c.

18. Describe the difference between a crosscut saw and a ripsaw.

19. Explain why safety glasses are necessary when driving nails with a hammer.

20. What measurements are indicated in Figure 13–53?
   a.
   b.
   c.
   d.
   e.

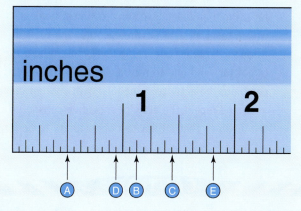

Figure 13–53

Activities

PURCHASING TOOLS

Among the first things you will do in construction is buy tools. Perhaps you already have a start. For this activity, assume you have $100 to buy hand tools. Using catalogs provided by your instructor, catalogs you find on your own, and research on the Internet, make a list of the tools you will buy. For each tool, give its specifications (what it is made of, its features, etc.) and its price. Make a separate list of the tools you plan to buy next.

USING A BUILDER'S LEVEL

The builder's level is used by many of the construction trades, because it is the surest way to measure elevations at various places. In this activity, you will follow the instructions given earlier in the chapter for using a builder's level. This can best be done with a team of at least three people, one to sight through the telescope of the level, one to hold the target rod, and one to record elevations. When you complete the activity, trade jobs with another teammate and repeat the activity, including setting up the level, so that all have an opportunity to perform all three jobs.

CAUTION

CAUTION: The builder's level can easily be ruined by rough handling. Treat it with care.

Equipment and Materials

- Builder's level
- Target rod (if a target rod is not available, a pole with a tape measure attached to one side can be substituted)
- Notepad and pencil
- 2 stakes
- Sledgehammer
- Claw hammer (to start stakes)

PROCEDURE

1. Drive two stakes in the ground about 10 or 15 feet apart. The depth to which you drive the stakes is not important, but they should be solidly in place. Do not drive them too deep since you will have to remove them when you are finished.
2. Following the instructions in this chapter, set up and level the builder's level about 20 feet away from the stakes. Have your instructor check your level when you have it ready for use.
3. Rest the target rod on the top of one stake, being careful to hold it plumb. If the stake is not plumb, your readings will be off (Fig. 13–54).
4. Focus the telescope on the target rod, and record the elevation. Be careful not to touch the tripod when you turn the level.
5. Rest the target rod on the other stake, and focus the level on the rod at this stake. Record this elevation.
6. Which stake is higher and by how much?

   Note: The higher stake will give a lower reading, because the distance from the top of the stake to the focal point is shorter.

7. Carefully, return the builder's level to its case and fold up the tripod.
8. Give the stakes a light tap with the sledgehammer to change the readings.
9. Switch jobs, and repeat steps 1 through 7 until all team members have had a chance to do all three jobs.

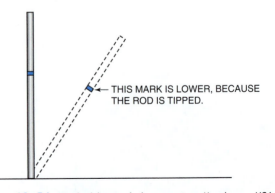

THIS MARK IS LOWER, BECAUSE THE ROD IS TIPPED.

Figure 13–54 *Used with permission, courtesy Handyman USA LLC.*

BUILDING A SAWHORSE

Equipment and Materials

- One 2 × 6 at least 33 inches long
- Four 2 × 4 at least 9 feet long
- Approximately 8 feet of 1 × 4
- Two pieces 1/2- or 3/8-inch plywood 12 inches × 12 inches
- 12d common nails (see Chapter 15 for nail sizes)
- 6d common nails (see Chapter 15 for nail sizes)
- Crosscut saw
- Rafter square or speed square
- Claw hammer
- Tape measure

PROCEDURE

1. Cut one piece of 2 × 6 33 inches long.
2. Using a square, mark each corner of the 2 × 6 where the 1½ × 3½ legs will attach (Fig. 13–55).

3. Using the crosscut saw to make the cross-grain cuts and the ripsaw to cut with the grain, cut out the notches for the legs.
4. Cut 4 pieces of 2 × 4 × 28 inches long, for the legs.

   Note: Always mark your cuts with a square before you start sawing. The results will be truer cuts and better-fitting pieces.

5. Use two 12d nails in each leg to attach the legs to the top. Let the legs extend about 1/2 inch above the top so that they can be sawed off flush later. It is important that all legs be the same length below the top so that they rest evenly on the floor. Do not drive the nails all the way home. The joint needs to be loose until all parts are in place.
6. If the plywood has not yet been cut into 12-inch squares, do that now.
7. Position the legs at one end 19 inches apart, and with them in that position, scribe the edges of the plywood square to indicate where it should be cut to shape. Marking both angled sides equal will ensure the legs are set at an equal angle.
8. Cut the first plywood end to the right shape, using a crosscut saw.
9. Using the end that has been cut to shape, trace the second end and cut it the same as the first.
10. Nail the plywood ends in place with 6d common nails (Fig. 13–56).

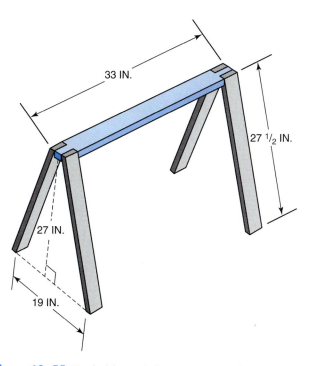

Figure 13–55 *Used with permission, courtesy Handyman USA LLC.*

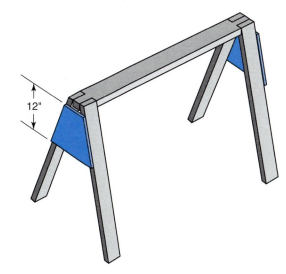

Figure 13–56 *Used with permission, courtesy Handyman USA LLC.*

Activities

11. Cut two pieces of 1 × 4 the same length as the top, 33 inches.

12. Using 6d nails, nail the 1 × 4 to the sides of the legs right under the edge of the plywood (Fig. 13–57). (Screws can be substituted for nails to make a stronger sawhorse, but driving all those screws by hand will be a chore. We shall discuss using an electric drill to drive screws in the next chapter.)

13. Hold a piece of 1 × 4 up to the end of the sawhorse, and scribe it to be cut to fit the end.

14. Nail the 1 × 4 on the ends.

15. Drive the 12d common nails in the tops of the legs home.

Note: If you build a second sawhorse, make the top only 31 inches long so that it can be stacked on top of the other.

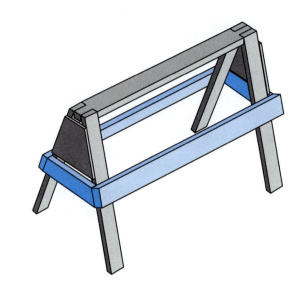

Figure 13–57 **Finished sawhorse.** *Used with permission, courtesy Handyman USA LLC.*

Power Tools— Selection, Use, and Care

After completing this chapter, the student should be able to:

- discuss basic specifications for common power tools.
- explain what operations are commonly done with each of the tools discussed.
- safely perform simple operations with common power tools.

Glossary of Terms

ampere the unit of measure for electric current. Also abbreviated amp. Many power tool motors are sized according to the amperage their motor draws.

ball bearing a style of bearing in which moving parts roll on steel balls.

chuck the part of a drill that holds the drill bit.

chuck key a special tool used to tighten a drill chuck.

combination blade a saw blade that can be used for ripping and crosscutting.

double-insulated a style of electric tool construction that shields the user from the electric parts of the tool.

plunge cut a cut made by plunging the saw in the middle of the work piece instead of cutting in from an outside edge.

pneumatic tools tools powered by compressed air.

polarized plug an electric plug having one prong wider than the other so that it can only be plugged into the receptacle one way.

reciprocate move back and forth in a straight line. Some saw blades reciprocate.

reversing switch a switch found on most electric drills that allows the user to reverse the direction of the drill.

roller bearing a style of bearing in which moving parts roll on small steel rollers.

shoe as used on power tools, the part of the tool that rests on the work piece.

Skil saw a term sometimes used to mean a portable circular saw. The first portable circular saw was made by Skil, and so some people call all portable circular saws Skil saws.

sleeve bearing a style of bearing in which moving parts ride on a smooth metal sleeve or tube.

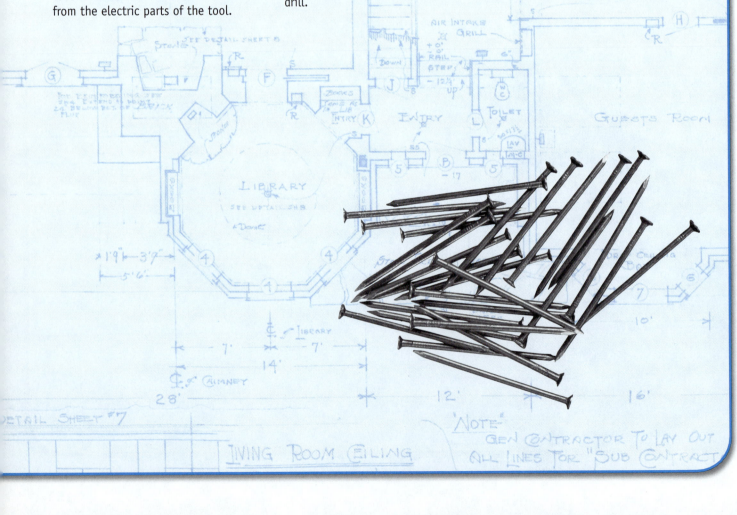

You will use power tools regardless of what building trade you work in. A power tool will often do the same job as a hand tool, but the power tool will do it more quickly and sometimes more accurately. Power tools can be dangerous, but by following all safety guidelines and using the tools properly, they are perfectly safe. Never use a power tool until you have been trained in its proper use. Quality power tools are expensive, and so it is important to know how to care for them.

There are a great many power tools that are not covered in this chapter. Those that are used primarily by only one trade have not been included, but they are usually discussed thoroughly in textbooks for the specific trade.

Power Tool Construction

Motors

The size of the motor on most power tools is specified by the amount of electric current the tool uses. Electric current is measured in **amperes** (abbreviated amps). A typical motor size for a circular saw might be 13 amps (Fig. 14–1). The higher the amperage, the higher the power output of the tool. Of course, higher amperage generally means more weight and greater cost.

Cordless Tools

Most portable power tools are available in cordless models. These tools have a rechargeable battery that provides electricity to run the tool. When not in use, the battery is placed on a charger. Good-quality cordless tools will deliver nearly as much power as corded tools and can usually be used for several hours without recharging. Be sure to have two batteries on hand so that one can be kept on the charger while the other is in use (Fig. 14–2). Cordless tools offer two advantages over corded tools: They are more portable, and

there is little chance of electric shock because they operate on lower voltage.

Insulation and Grounding

To protect the users of corded power tools from shock, many are **double-insulated.** Double-insulated power tools have two chassis. The inner chassis contains all the electrical parts. The outer chassis is completely insulated from the inner one. In case of an electrical malfunction, the outer chassis prevents voltage from reaching the operator.

Power tools should also be connected to an electrical ground. (See Chapter 3 for more information about electrical grounding.) Most power tools have a three-wire power cord. The third wire is attached to a grounding prong on the plug (Fig. 14–3). The ground connection in an electric outlet is connected to an electrical ground.

All corded tools made in the last 15 years will have a **polarized plug**—where one blade is wider than the other. This design ensures that the plug can only be inserted into

Figure 14–2 **Battery charger for cordless tool.**

Figure 14–1 **The motor label shows the amperage of the tool.**
Used with permission, courtesy of Makita U.S.A., Inc.

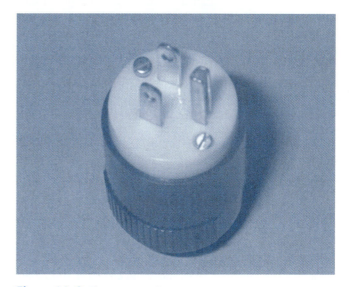

Figure 14–3 **Three-prong plug.**

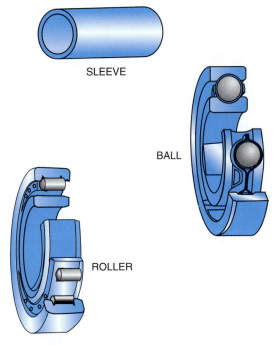

Figure 14–4 **Types of bearing used in power tools.**

the outlet one way to prevent electric shock. Do not attempt to force the plug to fit into an outlet it does not fit.

Pneumatic Tools

Pneumatic tools are those that use compressed air for their power. The most common pneumatic tools are nailers and staplers, but there are also pneumatic grinders, saws, drills, sanders, and wrenches. Pneumatic tools are sometimes lighter weight than their electric counterparts because they do not have electric motors. Of course, a pneumatic tool does require an air compressor and air hose.

Bearings

Moving parts cause friction, which consumes power and wears out parts. By putting bearings on rotating shafts, the friction is greatly reduced. Three types of bearing are used in power tools (Fig. 14–4). Sleeve bearings consist of a smooth surface on the inside of a short sleeve. Sleeve bearings have a shorter life expectancy than the other types, and they are not widely used in the best power tools. Roller bearings contain free-rolling, straight rollers that allow the shaft to roll in the support of the bearing. Ball bearings are a type of rolling bearing. Ball bearings are usually found where the bearing must resist end-to-end as well as side-to-side movement.

Power Tool Safety

There are specific safety rules that pertain to individual tools, but the following rules should be observed around all power tools:

- Use power tools only after you have received instruction in their use.
- Tools that are not double-insulated must be connected to an electrical ground.
- Wear appropriate protective clothing when operating power tools. Always wear safety glasses, and use hearing protection around tools that make loud noise. Do not wear loose clothes or jewelry, and do not wear your hair in a style that may cause your hair to get entangled in the tool.
- Keep all guards and protective devices in place.
- Do not use defective tools.
- Unplug corded tools and pneumatic tools when changing blades, bits, and attachments.
- Disconnect electric and pneumatic tools when they are left unattended.
- Pneumatic tools should be connected with a quick-disconnect coupler.
- Check to see that the cutting edge of the tool will have a clear path.
- Only plug corded tools into GFCI-protected outlets. (See Chapter 3 for information about GFCI protection.)
- Do not force the tool.
- Do not operate electric power tools near flammable liquids or in a dangerous atmosphere. Even cordless tools can ignite flammable vapors.

Portable Circular Saw

In 1928 a company named Skil invented a portable circular saw (Fig. 14–5). The Skil Saw saved a lot of time and made carpenters' work easier. Since then, many people have called all portable circular saws Skil saws. There are actually many manufacturers of portable circular saws.

The portable circular saw consists of a motor, a handle with a trigger switch, a shoe that tilts up to 45° and moves

Figure 14–5 **The first Skil saw.** *Courtesy of Robert Bosch Tool Corporation.*

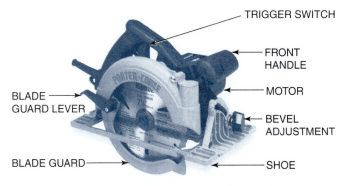

Figure 14–6 **Portable circular saw.** *Courtesy of Porter Cable.*

up and down to vary the depth of cut, a blade, a blade guard, and various other features (Fig. 14–6). (A **shoe** is the part of the tool that rests on the work piece.) Portable circular saws are sized by the diameter of their blade, with 7¼ inch being a popular size. Larger saws are available, but they are tiring to use for long periods of time. A smaller saw might not cut through a 2-inch piece of material at an angle.

Various blades are available for different operations (Fig. 14–7). The blade used most often is a **combination blade.** The combination blade can be used for crosscutting and ripping, the two most commonly performed operations.

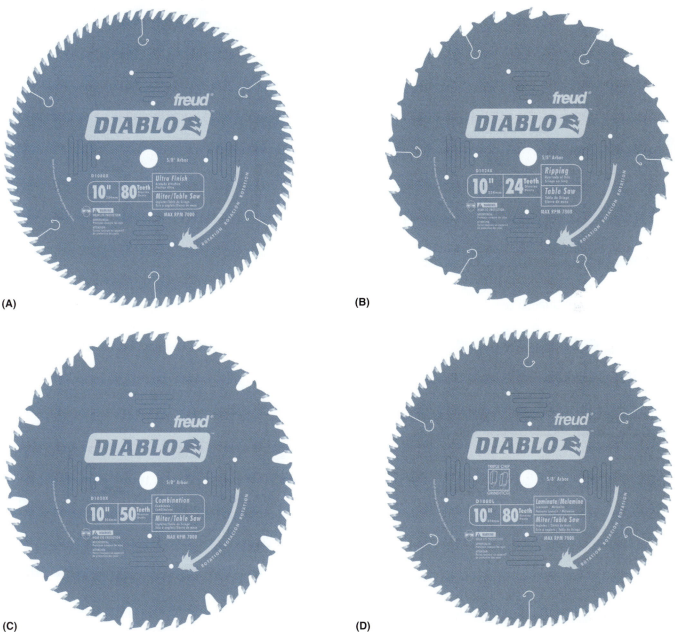

Figure 14–7 **Types of circular saw blades: (A) crosscut, (B) rip, (C) combination, and (D) laminate cutting blade.** *Courtesy of Freud®* TMM, Inc.

Inspection and Defects

Before you plug the saw in, inspect it to make sure it is in good condition and safe to use. Inspect the power cord to see that the insulation has not been broken, the plug is not broken or deformed, and the point where the cord attaches to the saw is tight and in good condition. Check to see that the blade is installed properly, with the teeth pointing up in front, and that it is the proper blade for the job you will do. Check the guard to see that it retracts completely. Are there any loose parts? The guard is the only part that should move. Do not use the saw if any of the above defects are present.

Crosscutting (Cutting Off a Piece of a Board)

Crosscutting is the most common use of a portable circular saw.

1. Check to see that a crosscut or combination blade is installed with the teeth pointing in the right direction (up in front).

CAUTION: Do not overtighten the blade nut. Most saws have a special washer under the nut that allows the blade to slip if it binds in the cut. This prevents the saw from kicking back out of the cut and injuring the operator.

2. Adjust the shoe so that the blade penetrates about 1/8 inch to 1/4 inch through the stock.

CAUTION: Do not set the blade to cut deeper than necessary. Keeping the blade inside the cut will shield much of it from the operator.

3. Check the blade guard to see that it works freely.
4. Rest the piece to be cut on sawhorses so that none of the portion to be cut off is supported (Fig. 14–8).

CAUTION: The stock being cut off should not be held or supported in any way. To do so might cause the blade to bind in the kerf. Let it fall free.

5. Mark a line on the stock where it is to be cut off, using a square and a sharp pencil.

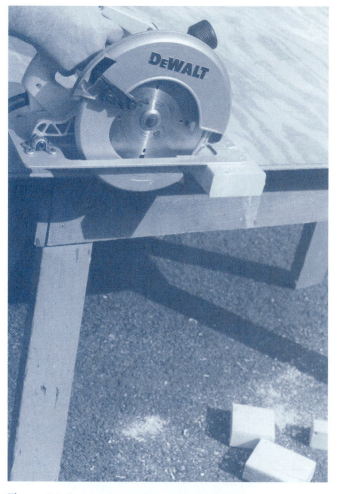

Figure 14–8 Saw cuts made over the end of the supports so that the waste will not bind the blade.

6. Plug the saw in.
7. Grasp the saw by the two handles and turn it on.
8. Rest the forward end of the shoe on the stock, and push it slowly but firmly into the stock. Avoid turning the saw from side to side. Keep the saw kerf (blade) on the scrap side of the pencil line.

CAUTION: If the saw binds, turn it off without trying to remove it from the kerf. With the saw off and unplugged, reposition the stock so that the saw can be easily removed. Start the cut over at the beginning.

9. After the cut is complete, release the trigger switch but hold the saw until the blade stops. Check to see that the blade guard has returned to cover the blade before setting the saw down.
10. Unplug the saw.

Ripping

Ripping is done in much the same way as crosscutting, with the following changes:

1. The blade must be either a combination or a rip blade. Usually a combination blade is used for ripping.
2. The board being ripped must be supported in such a way that the blade will not hit the sawhorse or other support (Fig. 14–9). This often means stopping part way through the cut to reposition the stock.

Making Plunge Cuts

It is sometimes necessary to make an internal cut—a **plunge cut**—such as for an opening in a floor or wall. To make these cuts, the saw must be plunged into the material.

1. Carefully lay out the cut to be made, paying particular attention to anything that might be on the other side of the material.
2. Adjust the saw for the proper depth of cut (1/8 inch to 1/4 inch through the material).

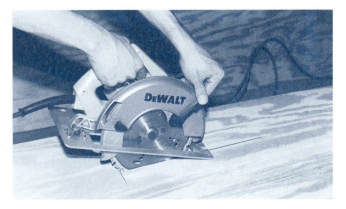

Figure 14–10 Making a plunge cut with a circular saw.

3. Hold the guard open, and tilt the saw up on the front of the shoe.
4. Move the saw so that the blade lines up with the place where you want the cut to be made.
5. Start the saw, and slowly lower the blade into the work (Fig. 14–10).
6. Follow the line carefully until the entire shoe rests on the material.
7. Release the guard so that it covers the blade.
8. Push the saw through the cut until you reach the corner of the opening.

CAUTION

CAUTION: Do not move the saw backward, as it may climb out of the cut, cause damage to the material, and injure the operator.

9. Release the trigger switch, and wait for the saw to stop turning before lifting it out of the kerf.

Saber Saw

The saber saw (Fig. 14–11) is widely used to make curved cuts and to saw small openings, such as for electric outlets. Some saws can be switched from straight up-and-down strokes to orbital (circular motion) cuts to provide the most effective cutting action for different materials. Many saber saws have a variable-speed switch, allowing the user to control how fast the blade moves. The shoe can be tilted to make bevel cuts. Many blades are available for cutting wood, metal, plastics, and other materials (Fig. 14–12). Blades with fewer teeth per inch cut faster but rougher. Blades with more teeth per inch cut slower but produce a smoother cut. Narrow blades are used to cut sharper curves, but wider blades are stronger and easier to use to cut straight lines.

Figure 14–9 Ripping with the stock positioned so that the blade will not touch the supports.

Figure 14–11 Saber saw.

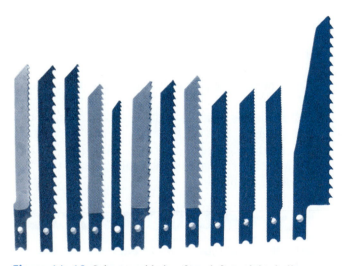

Figure 14–12 Saber saw blades, from left to right: hollow ground (smooth cuts in wood), fast cuts in wood, medium cuts in wood, flush cuts in wood, scroll cuts in wood, reverse teeth for fast cuts in plastic laminate, fast cuts in plastic laminates, medium cuts in wood and plastic, metal to 1/4" thick, aluminum and plastic, metal to 1/8" thick, fast cuts in plaster.

Size is not generally a consideration with saber saws. They all use the same blades, and they are used for light sawing, and so the power of the motor is not a factor.

Inspection and Defects

Inspect the saw for visible defects. Inspect the power cord to see that the insulation has not been broken, the plug is not broken or deformed, and the point where the cord attaches to the saw is tight and in good condition. Look for loose parts. There should be none.

Using a Saber Saw

1. Select the best blade for the type of cutting you will do.
2. Mark the cut with a pencil.
3. Wear safety glasses.

4. Hold the shoe of the saw firmly on the work. With the blade clear, turn on the saw. If it is a variable-speed saw, set the switch for the best speed for your work.
5. Push the saw into the work slowly, keeping the kerf on the waste side of the pencil line. Maintain firm downward pressure on the saw to reduce vibrations and improve cutting speed.
6. Turn the saw slowly as it advances to follow curves. Do not twist the saw without advancing. This will break the blade and will result in a rough cut. Do not force the saw into the work. Advance it only as fast as it will cut easily.

Making Plunge Cuts

Plunge cuts can be made with a saber saw in a manner similar to that used with a portable circular saw.

1. Mark the cut to be made with a pencil, checking to see that nothing behind the material will be in the path of the blade.
2. Tilt the saw up on the front edge of the shoe, with the blade in line with the pencil mark and clear of the work (Fig. 14–13).
3. Turn the switch on, holding the base steady. (Hint: If the saw has a variable-speed switch, a slow speed may work best to start a plunge cut.)
4. Gradually tilt the saw down until the blade penetrates the work and the base rests firmly on it.
5. Cut along the line, staying on the waste side, but as close as possible to the line until you reach the corner.
6. Back up about an inch (it is not necessary to turn off the saber saw to back up within the cut, but do not take it out of the cut), turn the corner by cutting a

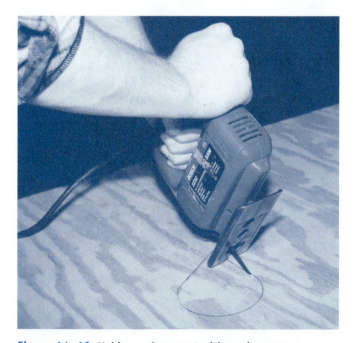

Figure 14–13 Making a plunge cut with a saber saw.

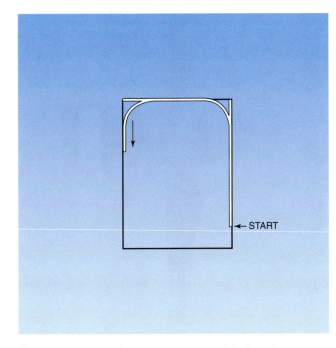

Figure 14–14 Cutting out an opening with the saber saw.

Figure 14–15 Reciprocating saw.

small arc, and cut along the other side and into the next corner (Fig. 14–14). Continue in this manner until all four sides have been cut.

7. Turn the saw around and cut in the opposite direction to remove the small pieces left in the corners.

Reciprocating Saw

The reciprocating saw (Fig. 14–15) is larger and more powerful than a saber saw, but has a blade that moves back and forth—**reciprocates**—like a saber saw. It is definitely a two-hand saw. A selector switch allows the user to select between a straight-line motion or an orbital (circular) motion. Most reciprocating saws will accommodate all the blades made for reciprocating saws, and so the only way to indicate the size is by the amperage of the motor.

CAUTION

CAUTION: Do not attempt to move the selector switch with the saw turned off. This can cause damage to the switch.

Blades are available for making rough cuts in wood, for making smoother cuts in wood, for cutting metal, for cutting plastics, and for doing demolition work (Fig. 14–16). Reciprocating saw blades are also available in different lengths. A reciprocating saw will cut faster than a saber saw or handsaw and can be used where a circular saw would be either unsafe or difficult to use. A reciprocating saw will not, however, cut the sharp curves that a saber saw can cut easily.

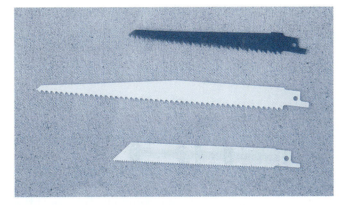

Figure 14–16 Reciprocating saw blades.

Inspection and Defects

Inspect the saw for visible defects. Inspect the power cord to see that the insulation has not been broken, the plug is not broken or deformed, and the point where the cord attaches to the saw is tight and in good condition. Look for loose parts. The only part that should be able to move is the base shoe, which moves to adjust to the angle at which the saw is being used.

Using a Reciprocating Saw

The reciprocating saw is a fairly safe tool to use. However, like any other power tool, it is necessary to know how to use it, and certain safety precautions should be observed.

1. Insert the right blade for the job in the saw, and tighten it according to the manufacturer's instructions.
2. Check to see that nothing will be in the path of the blade except that which you intend to cut. A reciprocating saw will make quick work of a pipe or electrical cable.

CAUTION

CAUTION: When checking to see what will be in the way behind your work piece, do not forget to consider the length of the stroke.

3. Ensure that the workpiece is held securely. The rapid reciprocating motion of the saw can really shake a loose workpiece!
4. Hold the saw comfortably in both hands, and pull the trigger switch to turn the saw on. Then move the selector switch to the desired position.
5. Place the shoe against the edge of the workpiece, and advance the saw into the cut slowly.

CAUTION

CAUTION: Do not twist the saw or attempt to cut curves. This will break the blade.

Drills

Many different sizes and types of power drills are used in construction. Some are classified as light-duty and others as heavy-duty drills. All power drills have a motor, handle(s), trigger switch, **reversing switch** (lets you reverse the direction of the drill), and chuck (Fig. 14–17). The **chuck** is the part of the drill that holds the drill bit. The size of the drill is specified according to the maximum size the chuck is designed to hold. Most light-duty drills are 1/4-inch or 3/8-inch drills. Heavy-duty drills are generally 3/8-inch or 1/2-inch drills. Light-duty drills are more comfortable to use than heavy-duty drills and often have more than enough power to do the job. Most construction workers use cordless drills where they can so they do not have to worry about a power supply and a power cord.

The part of a drill that actually does the cutting is called a drill bit. There are many types of drill bits, but the ones used most often are twist drill bits, masonry bits, and spade bits (Fig. 14–18). Some inexpensive twist drill bits are intended only for drilling in wood and other soft materials, but most can be used for wood, metal, and plastics. Drill bits come in sizes from only slightly larger than a human hair up to 1 inch or more. The twist drill bits you will use most will range from about 1/16 inch up to 1/2 inch. Spade bits are to be used only in wood, but they are made in larger sizes, ranging from 3/8 inch up to 2 inches or more. Larger holes, such as for locks in doors, are made with a hole saw (Fig. 14–19). Masonry bits are similar to twist drill bits, but they have a carbide cutting tip especially for drilling in concrete and masonry products. Screwdriver bits are also made especially for use in light-duty drills and power screwdrivers.

Figure 14–17 Portable power drills are available in several styles and sizes. *Courtesy of Porter Cable.*

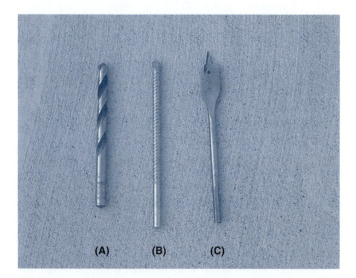

(A)   (B)   (C)

Figure 14–18 Common drill bits: (A) twist drill bit, (B) masonry bit, and (C) spade bit.

Figure 14–19 **Hole saw.**

Inspection and Defects

Inspect the power cord to see that the insulation has not been broken, the plug is not broken or deformed, and the point where the cord attaches to the saw is tight and in good condition. Check the chuck to make sure the jaws move uniformly as you tighten and loosen the chuck and that the jaws can close all the way.

Using a Power Drill

There are so many uses for a power drill that it is not possible to cover them all in this book. The following are general guidelines for using a power drill to drill a hole in wood or metal:

1. Make sure the workpiece is held securely. Use clamps, or temporarily nail it in place if necessary.

CAUTION

CAUTION: **Never attempt to hold a piece with your hands while you drill it.**

2. Check behind the workpiece to ensure that the bit will have a clear path.
3. Place the proper bit in the chuck, and hand-tighten the chuck. Most cordless and many other light-duty drills have keyless chucks. They can be fully tightened by hand. All heavy-duty and many light-duty drills require a **chuck key** (Fig. 14–20) to tighten the bit in the chuck. Make sure that the bit is centered between the three jaws of the chuck and that it is straight and centered when it is tightened.

CAUTION

CAUTION: **It is very dangerous to use a chuck key without first disconnecting the drill from the power source.**

Figure 14–20 **Chuck key.**

4. Check to see that the reversing switch is set in the forward position.
5. Make a small indentation or dimple where the center of the hole should be. This can be done with a center punch or, in wood, with a nail. This is to keep the drill bit from "walking" out of position when you start to drill.
6. Plug the drill in and pull the trigger switch. Use a medium to fast speed for wood and a low speed for metal.
7. Place the point of the drill on the indentation made in step 5 and apply moderate pressure. If you are drilling in metal, especially metal thicker than 1/16 inch, a drop or two of oil in the dimple will help the bit cut more smoothly and will help reduce overheating. Do not use oil on wood, as it will permanently stain the wood.

CAUTION

CAUTION: **Do not apply too much pressure, and do not let the drill tip to the side. Keep your drill in a straight perpendicular line to avoid breaking the bit.**

CAUTION

CAUTION: **If the bit binds in the cut, stop the drill, move the reversing switch to "reverse," and gradually turn the drill on to back the bit out of the hole.**

8. Be prepared for the bit to catch on the edge of the material as it passes through the far side of the hole. Try to maintain even pressure in a straight line, and keep the drill speed constant at this point.

9. If you are using a twist drill bit, keep the drill bit turning as you withdraw it from the hole. If you are using a spade bit or any other bit that is larger at the cutting end than the shank that passes through the hole, stop the drill completely and withdraw the stopped drill.

Using a Drill to Drive Screws

Many jobs can be done using a drill as a screwdriver or nut driver. Cordless drills are most convenient for this, but corded drills can be used, with the proper attachments. If you are using a corded drill, you will need a screwdriver attachment that has a clutch. This attachment allows the user to control how much torque is applied to the screw. There are power screw guns that look like an electric drill but have a chuck with a built-in clutch and a sleeve that stops the drill when it contacts the surface around the head of the screw. Most cordless drills have a clutch mechanism that the user can set to allow the drill to slip when it reaches a certain amount of torque (Fig. 14–21). If the chuck is set for full torque (normally used for drilling) when driving small screws, either the head of the screw will be twisted off or the screw will be buried deep in the work piece.

Figure 14–21 **Clutch on drill chuck.**

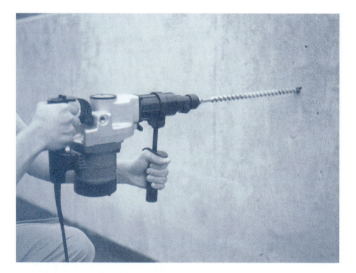

Figure 14–22 **Hammer drill.**

Hammer Drills

Hammer drills (Fig. 14–22) are similar to other drills except that they deliver as many as 50,000 hammer blows per minute to the drill point as the drill turns. The most popular sizes of hammer drills are 3/8 inch and 1/2 inch. The hammer blows improve the speed at which the drill will go through concrete and masonry.

Torches

Torches are used by plumbers and HVAC technicians to solder and braze copper pipe and fittings. There are several types of torches and at least three fuel gases that are commonly used for HVAC and plumbing. One of the most common types of torches is a small, self-contained unit that consists of a fuel cylinder with the torch head attached directly to the cylinder. One such torch is shown in Figure 14–23. The torch shown in the figure is a self-igniting torch that burns propane fuel. The torch mechanism includes an igniter that will light the torch when it is activated. The propane fuel that this torch burns is frequently used for soldering copper pipe, but it does not burn hot enough for brazing and welding. MAPP gas burns hotter than propane and can be used for brazing. Acetylene gas burns very hot, and it can be made to burn even hotter by mixing it with oxygen. Cutting and welding torches that have a double hose connecting the torch to two cylinders of gas are often oxyacetylene torches. Plumbers sometimes use acetylene for brazing and hard-soldering copper pipe.

Figure 14–23 Self-igniting torch. *Courtesy of BernzOmatic.*

Safety with Torches

Regardless of the type of torch or the fuel used, the following safety rules are important to know and observe:

- Know how to use the torch properly. You need to know how to control the torch flame and how to turn it off quickly should the need arise.
- Know the location of the nearest fire extinguisher and how to use it.
- Do not point the torch toward your face, any part of your body, or another person.
- Disconnect the cylinder when not in use.

- Be extra careful when using the torch outdoors on sunny days. The bright light can make it hard to see the flame, and wind can blow the flame back on the operator.
- Do not use a leaking, damaged, or malfunctioning torch.
- Never use a torch to strip lead-based paint indoors.
- Heating a surface may cause heat to be conducted to another part of the surface that is not intended to be heated. This can cause a fire, or the excess heat can cause pressure to build up to dangerous levels in an enclosed container or pipe.
- Always wear protective gloves when doing hot work.
- Never use a torch near combustible materials. Be especially careful around motor fuels.

Soldering a Pipe Fitting

1. Cut the pipe to length using a tubing cutter.
2. Ream to remove any metal burrs.
3. Dry-fit the pipe to ensure it goes squarely into the fitting.
4. Use steel wool or an emery cloth to thoroughly clean both the outside of the pipe and the inside of the fitting so it is as bright as a new penny.
5. Do not touch the cleaned area with your fingers. Immediately apply flux (soldering flux is a paste-like cleaner for this purpose) to these areas and insert the pipe into the fitting.
6. Secure the pipe and fitting so they cannot move during the soldering process.
7. Direct the flame at the fitting until the metal becomes hot enough to melt solder at contact.
8. Apply solder to the joint, feeding the solder along the joint and slightly behind the torch tip. Capillary action will draw the solder into the joint.
9. Continue heating until the solder flows smoothly around the joint and fills it completely. When the joint is full, the solder will form a tiny bead around the joint.
10. After you have finished soldering, clear away all flux residue around the joint with a clean, damp cloth. Let the joint cool by itself. Dipping it in water to cool will cause the joint to crack.

Review Questions

1 What are the 12 rules for use of power tools given in this chapter? Write these rules in your own words.

2 For each of the tools listed below, explain how the size is specified and describe one job that might be done with that tool.

- Combination circular saw blade:
    Size
    Use
- Circular saw:
    Size
    Use
- Saber saw:
    Size
    Use

- Reciprocating saw:
    Size
    Use
- Cordless drill:
    Size
    Use
- Twist drill bit:
    Size
    Use
- Hammer drill:
    Size
    Use

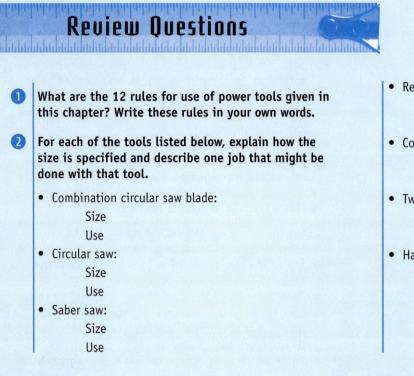

Activities

GROUND FAULT CIRCUIT INTERRUPTERS

Ground fault circuit interrupters are always used with corded power tools. After completing this activity, you should understand the operation of a ground fault circuit interrupter and appreciate how it can save lives.

Equipment and Materials

- 120-volt, 15-ampere ground fault circuit interrupter receptacle
- Two AC power cords (three-prong plug and three-conductor cable, minimum of 14-gauge wire)
- 25-watt lightbulb
- Lightbulb socket
- 12,000-ohm, 2-watt resistor
- Two alligator-to-alligator clip leads

INTRODUCTION

Ground fault circuit interrupters (GFCIs) are designed to sense an imbalance in the electric current flowing in the hot and neutral conductors. If an imbalance of greater than 0.006 ampere is detected, the GFCI will trip, removing power from the circuit. In this activity, you will use a GFCI receptacle to supply power to a lightbulb. You will use a resistor to simulate a fault that will cause an imbalance in the currents in the hot and neutral conductors.

PROCEDURE

1. Connect the circuit as shown in Figure 14–24.
2. Have your circuit verified by your instructor.

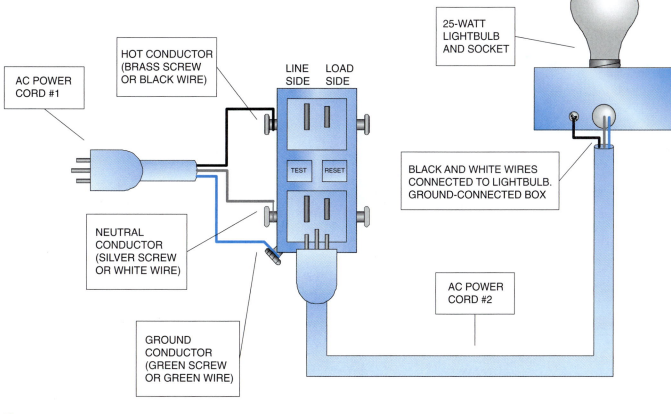

Figure 14–24

Activities

3. Apply power to your circuit by plugging AC power cord #1 into an energized 120-volt receptacle.
4. Verify that the lightbulb is lit.
5. Remove power from your circuit by unplugging AC power cord #1 from the 120-volt receptacle.
6. Connect the circuit as shown in Figure 14–25.
7. Have your circuit verified by your instructor.
8. Apply power to your circuit by plugging AC power cord #1 into an energized 120-volt receptacle.
9. Describe what happened.

ANALYSIS

1. Under normal conditions, does a GFCI behave much the same as a standard 120-volt, 15-ampere receptacle?
2. How is a GFCI different from a standard receptacle?
3. Do you feel it would be safer to use a standard receptacle or a GFCI receptacle? Justify your reasoning.
4. What are some locations that would be ideal for a GFCI in your home?

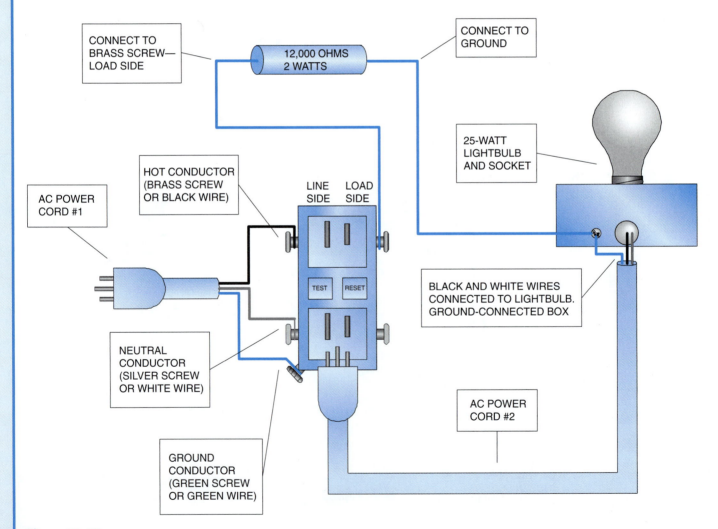

CONNECT TO BRASS SCREW— LOAD SIDE

12,000 OHMS 2 WATTS

CONNECT TO GROUND

25-WATT LIGHTBULB AND SOCKET

HOT CONDUCTOR (BRASS SCREW OR BLACK WIRE)

LINE SIDE   LOAD SIDE

AC POWER CORD #1

TEST   RESET

BLACK AND WHITE WIRES CONNECTED TO LIGHTBULB. GROUND-CONNECTED BOX

NEUTRAL CONDUCTOR (SILVER SCREW OR WHITE WIRE)

GROUND CONDUCTOR (GREEN SCREW OR GREEN WIRE)

AC POWER CORD #2

Figure 14–25

POWER TOOL FEATURES AND CONSTRUCTION

This activity will require you to have access to a computer and the Internet. Choose one power tool type for the activity, and find advertisements and specifications for at least three different manufacturers of that tool. Using that information, prepare a comparison chart for the three tools. The chart shown in Figure 14–26 should be used as a model for your comparison.

MAKING A TOTE CARRIER

Equipment and Materials

- Circular saw
- Drill
- 1-inch spade bit
- Phillips screwdriver bit
- Saber saw
- Square
- Tape measure
- Compass
- Pencil
- 3/4-in. × 2½-in. × 24-in. hardwood lumber (handle)
- 3/4-in. × 9¼-in. × 24-in. softwood lumber (bottom)
- 2 pieces 3/4-in. × 6½-in. × 25½-in. softwood lumber (sides)
- 1 piece 3/4-in. × 9¼-in. × 21½-in. softwood lumber (ends)
- 2-in. deck screws
- Glue

PROCEDURE

1. Use the circular saw to cut pieces of wood to the lengths described in the list of equipment and materials above.

   Note: The tote carrier has been designed to use pieces that are standard widths of softwood lumber. The only exception is the handle, which will be cut out with a saber saw. None of the pieces need to be ripped.

End Pieces

2. Use a square to ensure that the ends of the 21½-inch piece of wood are square.
3. Measure 10¼ inches from each end, and mark a line across the board. Measure 4⅝ inches along this line from the edge to find point A in Figure 14–27.
4. Use a compass to scribe a 1-inch-diameter circle with its center at point A.
5. Measure 6¼ inches from the ends of the board to find points B and C.

Feature	Acme Model 1515	ProTool Model 42	Tough Tools #145
Price			
Size			
Attachments included			
xxx			
yyy			

Figure 14–26 Suggested layout for Power Tool Features activity.

Activities

6. Complete the layout of two duplicate end pieces.
7. With the piece held in a vise or on a benchtop with a clamp, use the saber saw to cut out the two end pieces.

Handle

8. Lay out the handle as shown in Figure 14–28.
9. Saw the outside shape of the handle with a saber saw.
10. Mark the locations for the two holes for the handle.
11. Drill a 1-inch hole at each of the locations marked in step 10.
12. Connect the outside edges of the holes with straight pencil lines (Fig. 14–29).
13. Saw out the remainder of the handle with a saber saw or coping saw.
14. Smooth all surfaces near the handle grip with #80 abrasive paper.

Sides and Bottom

15. Check all parts to see that they are the correct size and that all ends are square. If all of the softwood was cut to the right size and with square ends in step 1, no further preparation is necessary for the sides or bottom.

Assembly (Fig. 14–30)

16. Attach the ends to the bottom with glue and 3 deck screws in each end.
17. Attach the sides to the bottom with glue and 6 screws and to the ends with 2 screws.
18. Position the handle and check to see that it is square—its sides are perpendicular to the bottom. Fasten with 2 screws in each end. Depending on how hard the wood is that you used for the handle, it might be necessary to predrill the screw holes in the ends of the handle. If this is necessary, drill 1/8-inch holes in the end pieces where the screws will go, and then line the handle up and mark the locations of the screw holes with a small nail, through these holes. Use the 1/8-inch drill to predrill the handle.

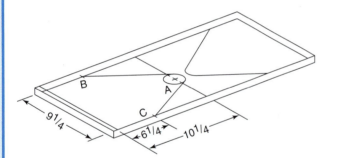

Figure 14–27 **End pieces.**

Figure 14–28 **Tote carrier handle.**

Figure 14–29 **Saw out handle along dotted lines.**

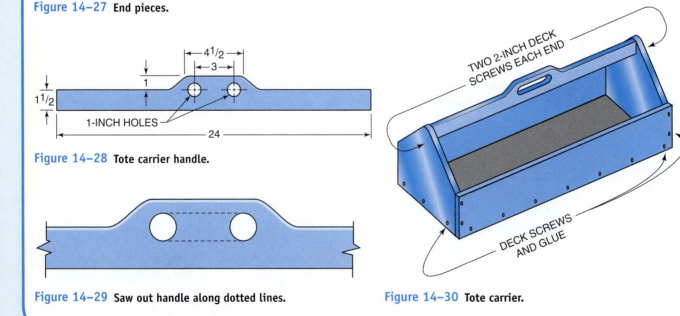

Figure 14–30 **Tote carrier.**

Chapter 15 Fasteners

OBJECTIVES

After completing this chapter, the student should be able to:

- ⊕ describe common fasteners.
- ⊕ select appropriate fastener types and sizes for common construction jobs.
- ⊕ install or use each of the common fastener types correctly.

Glossary of Terms

anchor a device that can be driven or set in concrete, masonry, or other material to provide a place to attach a bolt. There are several types of anchors.

box nail has a thin shank like a finishing nail, but a flat head like a common nail. Usually coated to prevent loosening.

brad a very short nail with a small head, used to fasten thin parts.

cap screw a small bolt, usually with a hexagonal head.

carriage bolt a large bolt for use in wood. Has a smooth oval head and a section of square shank right below the head.

clinching bending the protruding part of a nail over to make a permanent fastening.

common nail the most common type of nail. Has a heavy, smooth shank and a flat head.

deck screw similar to a drywall screw except it is more corrosion-resistant.

drywall screw a light-gauge screw with a Phillips or square-driven head, used for fastening drywall to framing. Drywall screws are not as strong or as corrosion-resistant as other screws.

duplex nails a common nail with two heads, so that one can be driven tight and the other is still exposed for removal.

finishing nail has a thin shank and small head that can be driven beneath the surface of the wood.

lag screw a large wood screw with either a square or hexagonal head.

penny size (abbreviated *d*) refers to the size of a nail. Nails are measured by an old system that used the number of pennies to purchase 100 nails of that size. The higher the penny size, the longer the nail.

pitch a measurement of the number of threads in 1 inch of a screw or bolt.

screw gauge a number representing the thickness of a screw. The higher the gauge, the thicker the screw.

sheet metal screw usually a self-tapping screw, used for fastening sheet metal.

stove bolt a small bolt with a round or flat head and fitted for a screwdriver.

Tapcon a concrete screw.

toenailing driving a nail at an angle into the face of one piece to hold it to another piece. Toenailing usually requires at least one nail on each side of the piece being toenailed.

surprising number of types and sizes of fasteners are available for use in construction. Some are intended for general-purpose use, and some are designed for one specific job. As a member of the building trades, you will be expected to be able to select the right fasteners for the work you do and to use those fasteners correctly. An improperly installed fastener often has far less holding strength than the same fastener properly installed.

Nails

Many kinds of nails can be used for a variety of fastening purposes. Several of the most common types of nails are pictured in Figure 15–1. Some of the nails shown here are intended for special purposes, and some are for general-purpose use.

For most nail types, the size is specified by **penny size.** The term "penny" (abbreviated *d*) was adopted in the early days of nail making. The penny size indicated the size of nails that could be purchased at that time at the rate of 100 nails for a given number of pennies. As you might imagine, a 10*d* nail in one town might have been a very different size from a 10*d* nail in another town. Today, the penny size of a nail indicates its length, regardless of what type of nail it is (Fig. 15–2). For example, 8*d* common nails, finishing nails, and box nails are different thicknesses, but they are all 2½ inches long. Roofing nails, staples, and brads are usually sized by wire gauge (the thickness of the wire used to make the nail) and length, not by penny size. Masonry nails are usually all the same gauge and are specified by length only.

Driving Nails

Nails can be driven in three ways: face nailing, toenailing, and clinching (Fig. 15–3). In face nailing, the nail is driven straight through one member and into the other. Face-nailed pieces are fairly easy to separate, by prying the pieces apart. **Toenailing** is used where it is not possible to use face nailing and where extra resistance to being pulled apart is

needed. A nail is driven into the side of one member at an angle, penetrating the adjacent surface and then into the other member. By toenailing from both sides, it becomes more difficult to separate the members, because the nails cannot be pulled straight out as the members part. **Clinching** refers to driving the nails through both members and then bending the points over to prevent withdrawal. Clinching is not often used in quality construction.

Use nails that are about three times as long as the thickness of the piece to be nailed. Select a hammer that is an appropriate size for the job to be done. A 20-ounce hammer would be appropriate for a 16*d* common nail, but not for a 4*d* finishing nail. A 7-ounce hammer might be okay for brads, but not for framing with large common nails. Start the nail with a couple of very light hammer blows and then drive it home. Hold the hammer firmly, but without squeezing it,

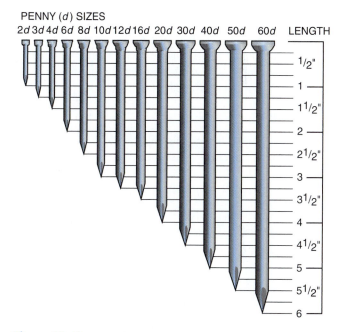

Figure 15–2 **Penny sizes of nails.**

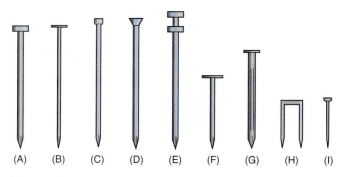

Figure 15–1 **Some of the most common kinds of nails: (A) common nail, (B) box nail, (C) finishing nail, (D) casing nail, (E) duplex nail, (F) roofing nail, (G) masonry nail, (H) staple, and (I) brad.**

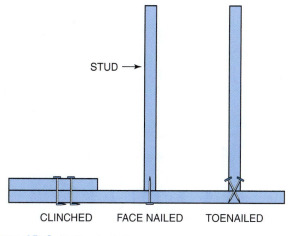

Figure 15–3 **Nailing techniques.**

near the end of the handle (Fig. 15-4). Keep your eyes focused on the head of the nail. With a little practice, the head of the hammer will consistently strike where your eyes are focused. If you strike the nail at an angle, it will bend. If the nail bends, pull it out and start over. It is much more difficult to drive a bent nail than a straight one.

Common Nails

Common nails are the most often used of all nails. **Common nails** have a flat head and a smooth shank. They are used for most applications where the special features of the other nail types are not needed.

Box Nails

Box nails are similar to common nails except they have a thinner shank and a thinner head. Because of their thin shank, box nails are less apt to split the wood used for boxes and crates. Also, box nails do not show up as readily under paint, because of their thin heads. Box nails are usually coated with a chemical that resists rusting and makes the nail more difficult to withdraw. Box nails bend more easily than common nails, and so it is especially important to hit them squarely when driving them.

Finishing Nails

Finishing nails have very small heads and somewhat thinner shanks than common nails. The small head of a finishing

nail can be driven below the surface of the wood and concealed with putty so that it is completely hidden. Finishing nails are used for installing trim and millwork where appearance is important. Because of their smaller heads, finishing nails have less holding power than common nails and box nails.

Casing Nails

Casing nails are kind of a cross between common nails and finishing nails. Casing nails have a small, cone-shaped head that can be driven below the surface of the wood. The shank of a casing nail is heavier than that of a finishing nail. Casing nails are designed for installing exterior door and window trim. In many cases this trim supports the full weight of the door or window, and so the nails must be strong.

Duplex Nails

Duplex nails have double (or duplex) heads. They are used for temporary structures, such as locally built scaffolds. The lower head ensures that the piece is nailed tightly. The upper head remains free of the wood, and so it is easy to pry the nail out when the structure is dismantled.

Roofing Nails

Roofing nails have very large heads so that asphalt and fiberglass shingles will not tear at the nail head. Unlike the nails described so far, roofing nails are specified by their length, not by penny size. Roofing nails are usually galvanized (a process of chemically coating steel with a combination of zinc and tin) so that they will not rust in exterior use. Roofing nails are often useful for other applications where the large head and the galvanized coating are helpful.

Masonry Nails

Masonry nails (Fig. 15-5) may be cut nails or wire nails. All masonry nails are made of hardened steel so that they will not bend as they are hammered into concrete or masonry. It is especially important to wear safety glasses when driving

Figure 15-4 Grasp the hammer near the end of the handle.

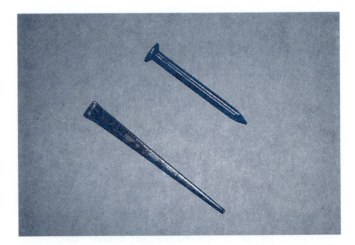

Figure 15-5 Masonry nails.

masonry nails, because they are brittle and may break when struck with a hammer.

Brads

Wire **brads** are very small nails shaped like finishing nails. They are used for finish work where a finishing nail would be too large.

Screws

Screws are widely used in construction where extra holding power is required. Screws are also used where the parts might need to be disassembled. For example, screws are used in door hinges, because the movement of the door might cause nails to work loose and because it is sometimes necessary to remove door hinges.

Screws are designated by their type, length, and gauge— a **screw gauge** is a number indicating the diameter of the shank. The type of screw usually determines the shape of the head and the pitch and/or depth of the threads. **Pitch** refers to the number of threads in 1 inch of a screw or bolt. The most common head shapes are flat, round, pan, oval, bugle, and hex (Fig. 15–6). There are also several types of drives or slots (Fig. 15–7).

This text describes the most common types of screws (Fig. 15–8), but there are many others.

Drywall Screws

Drywall screws generally come with Phillips drive, although there are some square recess drive types. Drywall screws are sometimes called Twinfast screws. Drywall screws

have a bugle head, and they range in length from 1 inch to 3 inches. Drywall screws are not made for exterior use and should not be used for decking or other exterior applications.

Particleboard Screws and Deck Screws

Particleboard screws and **deck screws** are similar to Twinfast drywall screws except that they are more corrosion-resistant. Particleboard screws are coated with black phosphate, and deck screws may be either galvanized or made of stainless steel. Both have a bugle head and are available with square-recessed drive or Phillips drive. Particleboard screws are the most common all-around wood screw for interior use, and deck screws are the most common for exterior use. Particleboard screws and deck screws can usually be driven without drilling a pilot hole first, but in hardwood it is always best to drill a pilot hole the same diameter as the screw shank to eliminate the problem of screws breaking.

Sheet Metal Screws

Sheet metal screws are used for fastening thin metal. Some sheet metal screws are self-tapping, which means the screw cuts its own thread in the hole as it is driven. Another type of sheet metal screw, called a tek screw, is self-drilling (Fig. 15–9). A tek screw drills its own pilot hole in the sheet metal as the screw is driven. This type of screw is used to fasten metal framing. Sheet metal screws are available in lengths from 1/4 inch up to several inches and in gauges from 6 gauge to 12 gauge.

CAUTION

CAUTION: Use care when drilling and handling sheet metal. The cut edges and fine scraps that are produced can be very sharp. Do not try to wipe scraps from the surface with your bare hand.

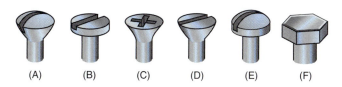

Figure 15–6 Screw head shapes: **(A)** oval, **(B)** pan, **(C)** bugle, **(D)** flat, **(E)** round, and **(F)** hex.

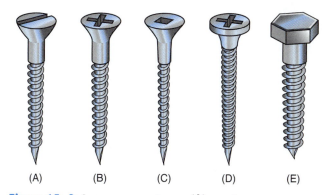

Figure 15–8 Common screw types: **(A)** wood screw, **(B)** Twinfast drywall screw, **(C)** particleboard screw, **(D)** pan-head sheet metal screw, and **(E)** lag screw.

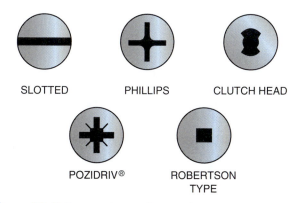

SLOTTED   PHILLIPS   CLUTCH HEAD

POZIDRIV®   ROBERTSON TYPE

Figure 15–7 Common types of screw slots.

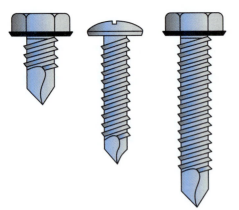

Figure 15–9 The point of a tek screw has a cutting edge that drills a hole as the screw is driven.

Figure 15–10 Cap screw.

Lag Screws

Lag screws are large wood screws with square or hex heads. Lag screws are turned with a wrench. These screws have great holding power and can be used to join timbers where a bolt cannot be used. Lag screws are sized by diameter and length. Diameters range from 1/4 inch to 1 inch. Lengths range from 1 inch to 12 inches. Two holes must be drilled for a lag screw. A shank hole the same diameter as the lag screw shank must be drilled through the top piece. The lower piece must have a pilot hole that is enough smaller than the shank to allow the screw to cut full threads in the wood. A flat washer should be placed under the head of the lag screw to prevent the head from digging into the wood.

Bolts

Bolts have high holding strength, and they allow for disassembly of parts that have been joined with bolts. Bolts are specified by type, diameter, and length. Most bolts are available in a variety of materials and coatings. Bare steel bolts are the least expensive, but they will rust if used outdoors. Most bolts that are to be used where they will be exposed to moisture are galvanized, but stainless-steel bolts are available for special applications.

Three types of bolts are widely used in construction: cap screws, stove bolts, and carriage bolts.

Cap Screws

Cap screws (Fig. 15–10) are actually a type of bolt. Some people call them machine bolts, but machine bolts are made especially for use on machines. Cap screws are available with hex heads, slotted heads for straight-slot screwdrivers, Phillips heads, and internal square drive or Allen drive heads. The smallest cap screws are sized according to the gauge number and thread pitch. The table in Figure 15–11

Gauge Number	Diameter in Inches
0	.060
1	.073
2	.086
3	.099
4	.112
5	.125
6	.138
8	.164
10	.190
12	.216

Figure 15–11 Number-size cap screws.

shows the sizes of number-size cap screws. Fractional-size cap screws range from 1/4 inch to 1 inch.

Cap screws are available with either Unified National Coarse (UNC) or Unified National Fine (UNF) threads. For example, a 10-32 cap screw is a 10-gauge screw with 32 threads per inch. That would be a UNF thread. The same cap screw with a UNC thread would be a 10-24 and would have 24 threads per inch. UNC is the most common, but it is important to remember that UNF threads exist. When selecting a nut for a cap screw, the threads must match. If the nut does not thread onto the bolt easily, it might be the wrong thread pitch, or the threads of either the bolt or the nut might be damaged.

Stove Bolts

Stove bolts are small bolts with either round or flat heads, and they are threaded all the way to the head

Figure 15–12 Stove bolt.

Figure 15–13 Carriage bolt.

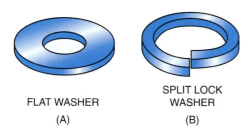

FLAT WASHER
(A)

SPLIT LOCK
WASHER
(B)

Figure 15–14 Washers: **(A)** flat washer and **(B)** split lock washer.

(Fig. 15–12). Stove bolts are generally made of low-grade steel and have a straight-slot or Phillips head. They usually have UNC threads and are supplied with hex or square nuts. Stove bolts are commonly used to join sheet metal parts, because they are threaded all the way to the head and strength is not usually a major concern with sheet metal fastening.

Carriage Bolts

A **carriage bolt** (Fig. 15–13) has a smooth round head with a square section right below the head. Carriage bolts are for use in wood. The square section of the bolt grips the wood and prevents the carriage bolt from turning as the nut is tightened with a wrench. A flat washer should be used under the nut, but no washer is used under the head of a carriage bolt. Carriage bolts are available from 1/4-inch to 3/4-inch diameter and in lengths from 3/4 inch to 16 inches. Most carriage bolts are zinc-plated or galvanized for exterior use.

Washers

There are several types of washers, but flat washers and lock washers are the most common (Fig. 15–14). Flat washers are used to protect the surface of the part from being damaged by the bolt head or nut. Washers are not used under the head of a carriage bolt, however. Split lock washers are used to keep the nut under tension and prevent it from loosening. When a lock washer is used, a flat washer

is usually placed under it to protect the surface of the part. Lock washers are not generally used on wood, because the wood itself is compressed a little and keeps the bolt under tension.

Anchors

Anchors are used to provide a means of attaching parts to concrete and masonry surfaces or to hollow walls. There are hundreds of types of anchors, and so only the most common ones are discussed here.

Wedge Anchors

Heavy-duty anchors are used to install machinery, hand rails, and anything that will put the anchor under a high load. A popular heavy-duty anchor is the wedge anchor (Fig. 15–15). To install a wedge anchor, use a masonry drill bit to drill a hole the same size as the anchor. Note: It is sometimes easiest to drill the part to be anchored and the concrete or masonry at the same time. This ensures that the holes line up perfectly. The hole should be deeper than the length of the anchor. Clean the hole out well enough to let you drive the anchor far enough into the hole so that at least six threads are below the top surface. Drive the anchor into the hole, position the piece to be anchored, and tighten the hex nut with a wrench.

Sleeve Anchors

The sleeve anchor (Fig. 15–16) is another popular type of anchor. The drill size is the same as the anchor size. Insert the sleeve anchor through the part to be anchored and into the hole in the concrete or masonry. Make sure the anchor is in firm contact with the piece to be anchored, and tighten it with a wrench. As the anchor tightens, the sleeve expands against the inside of the hole.

Split Fast Anchors

A split fast anchor (Fig. 15–17) is a medium-duty anchor. It is one-piece steel with two sheared, expanded halves at the bottom. When driven, these halves are compressed, and they exert a tremendous amount of pressure on the inside of the hole. Split fast anchors are available as round head or flat head styles.

Double Expansion Anchor

Double expansion anchors (Fig. 15–18) are used with cap screws. Drill a hole of the recommended diameter to a depth equal to the length of the anchor. Place the anchor in the hole, flush with or slightly below the surface. Position the part to be fastened, and bolt it into place, with a flat washer under the head. Once fastened, the part can be unbolted if desired.

DRILL - SIMPLY DRILL A HOLE THE SAME DIAMETER AS THE ANCHOR. DO NOT WORRY ABOUT DRILLING TOO DEEP BECAUSE THE ANCHOR WORKS IN A "BOTTOMLESS HOLE." YOU CAN DRILL INTO THE CONCRETE WITH THE LOAD POSITIONED IN PLACE; SIMPLY DRILL THROUGH THE PRE-DRILLED MOUNTING HOLES.

INSERT - DRIVE THE ANCHOR FAR ENOUGH INTO THE HOLE SO THAT AT LEAST SIX THREADS ARE BELOW THE TOP SURFACE OF THE FIXTURE.

ANCHOR - MERELY TIGHTEN THE NUT. RESISTANCE WILL INCREASE RAPIDLY AFTER THE THIRD OR FOURTH COMPLETE TURN.

Figure 15–15 Wedge anchor. *Courtesy of US Anchor, Pompano Beach, FL.*

Lag Shields

The lag shield (Fig. 15–19) is used with a lag screw. It is inserted into a hole of the recommended diameter and at least 1/2 inch deeper than the length of the anchor. The

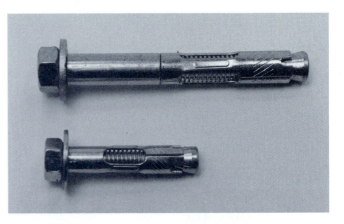

Figure 15–16 Sleeve anchor.

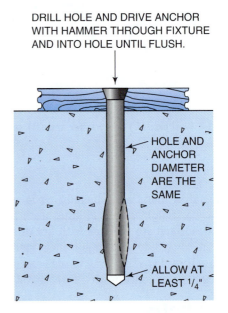

DRILL HOLE AND DRIVE ANCHOR WITH HAMMER THROUGH FIXTURE AND INTO HOLE UNTIL FLUSH.

HOLE AND ANCHOR DIAMETER ARE THE SAME

ALLOW AT LEAST 1/4"

Figure 15–17 Split fast anchor.

length of the lag screw should be equal to the thickness of the part to be fastened plus the length of the anchor plus 1/4 inch. The tip of the lag screw must project slightly beyond the bottom of the anchor to ensure that the anchor is fully expanded.

Tapcons

A **Tapcon** is a concrete screw (Fig. 15–20). Tapcons are widely used because they eliminate the need for a separate anchor and fastener. They use special threads to cut into the sides of a properly sized hole in concrete or masonry. The size of the hole is important, and so Tapcons come prepackaged with the right-size drill bit. To determine the proper-length Tapcon anchor for an application, combine the thickness of the part to be attached with the desired depth into the masonry materials. It is recommended that a minimum of 1-inch depth and a maximum of 1¾-inch depth into

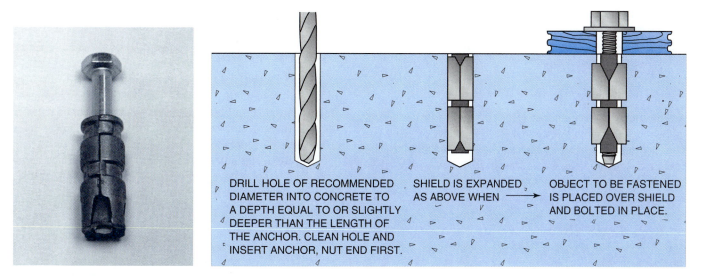

DRILL HOLE OF RECOMMENDED DIAMETER INTO CONCRETE TO A DEPTH EQUAL TO OR SLIGHTLY DEEPER THAN THE LENGTH OF THE ANCHOR. CLEAN HOLE AND INSERT ANCHOR, NUT END FIRST.

SHIELD IS EXPANDED AS ABOVE WHEN ———→

OBJECT TO BE FASTENED IS PLACED OVER SHIELD AND BOLTED IN PLACE.

Figure 15–18 **Double-expansion anchor.** *Courtesy of US Anchor, Pompano Beach, FL.*

Figure 15–19 **Lag shield.**

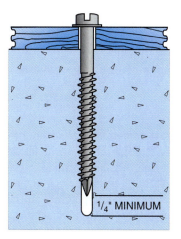

¼" MINIMUM

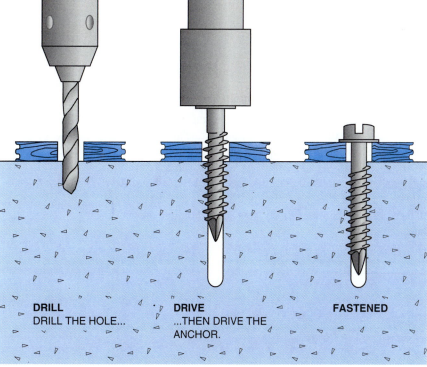

DRILL
DRILL THE HOLE...

DRIVE
...THEN DRIVE THE ANCHOR.

FASTENED

Figure 15–20 **Tapcon.**

the masonry be used when determining fastener length. Tapcon fasteners come in two diameters, 1/4 inch and 3/16 inch. The diameter of the fastener and the depth of embedment affect pull-out strength.

Hammer-Drive Anchors

The hammer-drive anchor (Fig. 15–21) has either a zinc alloy body or a nylon body and a steel expander pin. The anchor is inserted through the part to be fastened and into the masonry, and then the expander pin is driven with a hammer. These anchors are quick and easy to use, but are not as strong as the anchors discussed earlier.

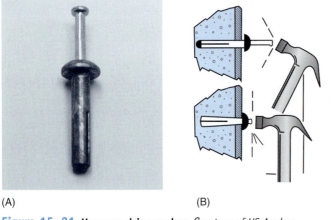

(A)   (B)

Figure 15–21 Hammer-drive anchor. *Courtesy of US Anchor, Pompano Beach, FL.*

Split Ribbed Plastic Anchors

Split ribbed plastic anchors (Fig. 15–22), also called conical screw anchors, are commonly used for attaching light loads. They are often sold as hollow-wall anchors (see the next section), because they will not hold much more weight than a hollow wallboard wall. This kind of anchor has ribs on the outside to keep it from slipping out of the hole as a screw is driven into it. As the screw penetrates the anchor, the split halves separate, increasing the grip in the hole. Another version of the same style anchor is made of two lead halves. Lead anchors offer slightly more holding power in masonry walls.

PLASTIC INSERT →

Figure 15–22 Split ribbed plastic anchor.

Hollow-Wall Fasteners

Toggle Bolts

Toggle bolts (Fig. 15–23) have spring-loaded wings. The wings are held against the sides of the threaded fastener as it is inserted through a hole in the wall. The wings must go beyond the wall surface material so that the spring can force them apart inside the wall. Then the screw is tightened, pulling the wings against the back of the wall-surface material. Toggle bolts have two disadvantages. One is that they require a hole in the wall that is large enough to fit the collapsed wings. The other is that if the screw is removed, the wings fall inside the wall.

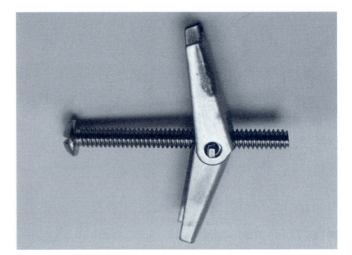

Figure 15–23 Toggle bolt.

Plastic Toggles

The plastic toggle (Fig. 15–24) has four legs attached to a body with a hole through it and fins on its sides. The fins prevent the toggle from turning in the hole. The legs collapse to allow insertion into the hole. As a sheet metal screw is turned into the body, the legs are expanded against the inner surface of the wall. Plastic toggles do not require as large a hole as do toggle bolts, and they are less expensive.

Molly Screws

Molly screws (Fig. 15–25) are sometimes called hollow-wall expansion anchors. They consist of an expandable sleeve, a machine screw, and a fiber washer. A collar at the surface of the wall acts like a washer, preventing the anchor from pulling into soft wall surfaces. Two sharp points on the collar prevent the sleeve from turning in the hole as the screw is turned. As the screw is tightened, the inner end of the sleeve is pulled toward the collar, expanding the sleeve.

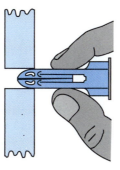

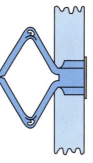

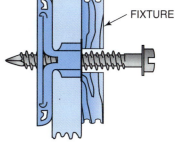

FIXTURE

SQUEEZE TOGGLE WINGS FLAT AND PUSH INTO HOLE DRILLED IN WALL.

TAP ANCHOR IN AND FLUSH WITH WALL.

PLACE FIXTURE OVER HOLE; INSERT SHEET METAL SCREW AND TIGHTEN.

Figure 15–24 **Plastic toggle.** *Courtesy of US Anchor, Pompano Beach, FL.*

Figure 15–25 **Molly screw.**

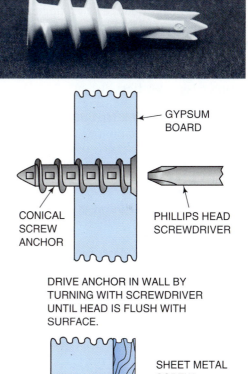

GYPSUM BOARD

CONICAL SCREW ANCHOR

PHILLIPS HEAD SCREWDRIVER

DRIVE ANCHOR IN WALL BY TURNING WITH SCREWDRIVER UNTIL HEAD IS FLUSH WITH SURFACE.

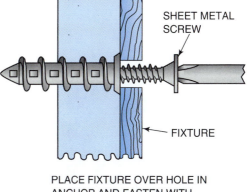

SHEET METAL SCREW

FIXTURE

PLACE FIXTURE OVER HOLE IN ANCHOR AND FASTEN WITH PROPER-SIZE SHEET METAL SCREW.

Figure 15–26 **E-Z Anchor.** *Courtesy of US Anchor, Pompano Beach, FL.*

Once the sleeve is expanded and is tight in the wall, the screw can be removed to position the piece to be attached to the wall. Some types of molly screws have a sharp point so that they can be driven into the wall material. Another type has a blunt end requiring that a hole be drilled first.

An item that has been installed or hung with a molly screw can be removed and refastened by removing the anchor screw without disturbing the sleeve. Molly screws are manufactured for different thicknesses of wallboard. Make sure you use the right size molly screw for the wall thickness.

E-Z Anchors

An E-Z Anchor (Fig. 15–26) has a sharp point and deep threads so that it can be driven into gypsum wallboard with a Phillips screwdriver. No drilling is required. When the anchor is driven fast, a sheet metal screw, deck screw, or drywall screw can be driven into the hole in the anchor.

Review Questions

Select the most appropriate answer where applicable.

1 What type of nails would be most appropriate for joining 2×4s?

2 What type of nails would be most appropriate for mounting a door frame in its opening?

3 What type of nails would be most appropriate for applying roof shingles?

4 What type of nails would be most appropriate for installing finished molding in a room?

5 What type of nails would be most appropriate for fastening furring (wood nailing strips) to a brick wall?

6 What kind of material or coating would be used to prevent a lag screw from rusting?

7 What does the abbreviation UNC mean when used in the specification for a cap screw?

8 What can be used to prevent a cap screw from vibrating loose when used on an air compressor?

9 What is the difference between a sheet metal screw and a cap screw?

10 What type of fastener described in this chapter would be used to bolt a sheet metal cover together on a heat pump?

For questions 11 through 16, match the fasteners in Column I with the uses and descriptions in Column II.

Column I	Column II
11 Tapcon	a. Plastic anchor for use in hollow walls. Can be driven with a screwdriver. No drilling required.
12 Wedge anchor	
13 Molly screw	
14 Lag shield	b. Lead anchor used in concrete to hold large screws.
15 Toggle bolt	
16 E-Z Anchor	c. Used in hollow walls. Attachment can be unfastened and refastened.
	d. Screw that can be driven directly into concrete.
	e. Used in hollow walls, where the part will cover a large hole and will not need to be unfastened and refastened.
	f. Used where maximum holding strength is needed in concrete and masonry.

17 Which is longer, a 6*d* finishing nail or a 10*d* casing nail?

18 Which has a larger diameter, a 10-32 cap screw or a 1/4-inch cap screw?

19 What is the greatest safety hazard when using concrete nails?

20 What keeps a carriage bolt from turning in its hole as the nut is tightened on it?

Activities

ALL ABOUT FASTENERS AND ANCHORS

There are many more types of fasteners than it would be possible to describe in this textbook, and there are more uses for fasteners and anchors in construction than also can be described here. In this activity you will find information about other types of fasteners and share that information with your class. The Internet is an excellent place to find information about many things, and this activity will give you practice in doing that.

Equipment and Materials

- Access to a computer and the Internet
- Hardware/fastener catalogs
- Poster board
- Tape, paste, markers, and other poster materials

PROCEDURE

Find as much information as you can about nails, screws, bolts, and anchors. Look for information on holding power, sizes, materials and finishes, prices, and the way to use the fasteners and anchors. Categorize your information so that you can present it in a well-organized display on poster board. Make a 5-minute presentation to your class about what you found. Your information will have to be planned well, and your presentation will have to be practiced so that you can present the most important information within the time allowance.

Hint: Most inexperienced presenters tend to go over their time allowance because they are not well organized and they wander off the main topics. Practice is the best way to avoid this pitfall.

FASTENERS OR ANCHOR DISPLAY

The best way to learn about a new item is to see it, handle it, and try out its features. Most of the hardware discussed in this chapter is intended for use in concrete and masonry, and it is not realistic for you to try out the fasteners in their intended materials.

In this activity you will obtain samples of several types of hardware and handle each to make a display describing its use.

Your instructor may assign you to work in pairs or groups of three on this activity. If so, everyone should be involved in collecting samples and finding information, and everyone should work on the display.

Equipment and Materials

- 1/4-inch plywood or other display backing
- Collection of fasteners or anchors
- Black marker
- Heavy epoxy adhesive or other means of attaching hardware to display board

PROCEDURE

1. Decide whether you will make a display of fasteners (nails, screws, and bolts) or anchors, including hollow-wall fasteners.
2. Collect as many samples as possible of different types of fasteners or anchors, and get information on how they are used. Much of that information might be in this chapter.
3. Make a display of the fasteners or anchors you collected, labeling each with its name and its purpose. Plan your display so that you can show sizes and types of materials for as many as possible. Arrange all the parts of your display without fastening anything permanently in place at first. Once you have decided on the final arrangement of your display, attach the fasteners or anchors to the display board with epoxy adhesive.

CAUTION

CAUTION: Read the instructions on the epoxy before you start mixing it. Wear appropriate protective equipment. Most epoxy can be removed by washing with soap and water until it begins to set, after which it cannot be removed at all. Spread a drop cloth or papers over the work area before you start mixing the epoxy.

Chapter 16 Rigging

OBJECTIVES

After completing this chapter, the student should be able to:

- safely rig a roof truss for hoisting by a crane.
- use standard hand signals for directing a crane operator.
- recognize common defects and safety hazards in rigging equipment.

Glossary of Terms

lay the direction in which the strands are wound around the core of a wire rope. Also the distance along a rope in which a strand makes a complete turn around the core.

rated capacity the amount of weight the manufacturer has specified for the maximum load on a sling.

tag line a light line attached to a load to control it by hand.

thimble a steel insert in the eye of a wire rope sling to prevent kinking and wear.

wire rope a rope made from strands of wire wrapped around a core in a particular way. Wire rope is sometimes incorrectly called cable.

wire rope clip a clamping device used to hold two pieces of wire rope to form an eye.

Rigging is the term used to describe the use of slings, ropes, straps, chains, and fittings to hoist heavy objects into place with a crane. In heavy construction, where much of the work is done at a considerable distance from the ground and where materials are often very heavy, cranes are used very frequently. The greatest use of cranes and rigging in residential construction is in lifting roof trusses into place. In some areas, air-conditioning equipment is commonly located on the roof, and so the equipment must be rigged for hoisting into place. Safety is a critical consideration in rigging. This chapter will introduce you to the equipment used in rigging, some key safety considerations, and basic rigging procedures. You will not be a rigger after only studying this chapter, but you will be able to work more safely around rigging.

Slings

The part of the rigging that attaches the load to the crane is called a sling. Practically all slings are made of wire rope, synthetic webbing, or steel alloy chains. Slings can be used in a variety of different ways (Fig. 16–1), depending on the load to be lifted.

All slings have a manufacturer's tag that shows:

- manufacturer's name.
- **rated capacity**—the amount of weight the manufacturer has specified for the maximum load—for the hitches or angles at which the sling should be used.
- type of material (type of material for synthetic slings, grade of steel and link size for chains, or size of wire rope).

Wear pads are used with wire rope and synthetic slings to protect the sling from sharp corners on the load.

Wire Rope

Construction

Wires are the basic building blocks of **wire rope.** They lay around a center in a specified pattern in one or more layers to form a strand. The strands lay around a core to form a wire rope (Fig. 16–2). Characteristics like fatigue resistance and

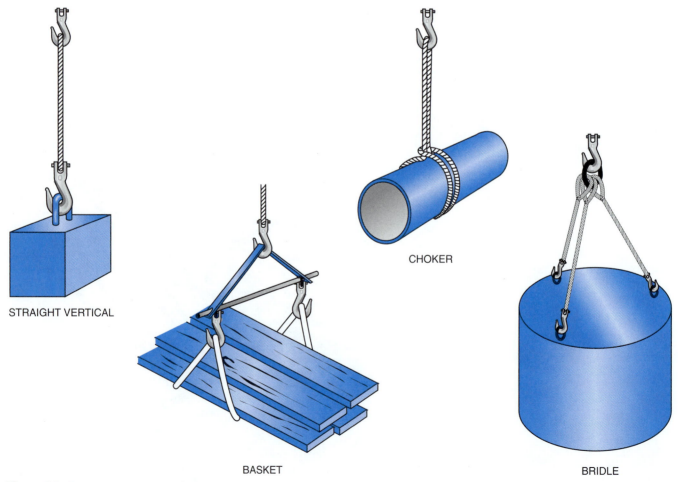

STRAIGHT VERTICAL

BASKET

CHOKER

BRIDLE

Figure 16–1 Four common sling hitches.

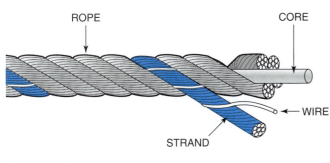

Figure 16–2 **Wire rope.**

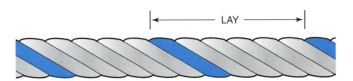

Figure 16–3 **Right lay.**

Figure 16–4 **Lay as a measurement.**

resistance to abrasion are directly affected by the design of the strands. In most strands with two or more layers of wires, inner layers support outer layers in such a manner that all wires may slide and adjust freely when the rope bends. As a general rule, a rope that has strands made up of a few large wires will be more abrasion-resistant and less fatigue-resistant than a rope of the same size made up of strands with many smaller wires.

Wire Rope Cores

The primary function of the rope's core is to serve as the foundation for the strands—to keep the rope round and the strands properly positioned during operation. The type of core has an effect upon the rope's performance. Three types of cores are most commonly used:

- *Fiber core.* Polypropylene (the kind of synthetic rope used to tow water skiers) is standard, but either natural fiber or other man-made fibers are available on special request.
- *Independent wire rope core.* Literally an independent wire rope with strands and a core, called IWRC. Most wire ropes made with steel core use an IWRC.
- *Strand core.* A strand made of wires. Typically, strand cores are used in utility cables only.

Lay and Rope Design

Lay has three meanings in rope design. The first two meanings describe the wire and strand positions in the rope. The third meaning is a length measurement used in manufacturing and inspection.

Meaning 1: The direction that strands lay in the rope— right or left. When you look down a rope, strands of a right lay rope go away from you to the right (Fig. 16–3). Left lay is the opposite. (It does not matter which direction you look.)

Meaning 2: The relationship between the direction that the strands lay in the rope and the direction that the wires lay in the strands. In appearance, wires in regular lay run straight down the length of the rope, and wires in lang lay appear to angle across

the rope. In regular lay, the wires are laid in the strand opposite the direction the strands lay in the rope. In lang lay, the wires are laid the same direction in the strand as the strands lay in the rope.

Meaning 3: The length of rope equal to one complete spiral of a strand around the rope core (Fig. 16–4). This is a measurement frequently used in wire rope inspection. Standards and regulations require removal when a certain number of broken wires per rope lay are found.

The lay of a rope affects its operational characteristics. Regular lay is more stable and more resistant to crushing than lang lay. Lang lay is more fatigue-resistant and abrasion-resistant.

Inspection of Wire Rope

All wire ropes will wear out eventually and gradually lose their ability to do the work for which they were intended. That is why periodic inspections are critical. Wire ropes should be thoroughly inspected at regular intervals. The longer the rope has been in service or the more severe the service, the more thoroughly and frequently it should be inspected. A record must be kept of each inspection (Fig. 16–5).

Inspections should be carried out by a person who has learned through special training what to look for and who knows how to judge the importance of any abnormal conditions he or she may discover. It is the inspector's responsibility to obtain and follow the proper inspection criteria for each application inspected.

Figure 16–6 shows what happens when a wire breaks under tensile load exceeding its strength. It is typically recognized by the "cup and cone" appearance at the point of failure. The necking down of the wire at the point of failure to form the cup and cone indicates failure has occurred while the wire retained its flexibility.

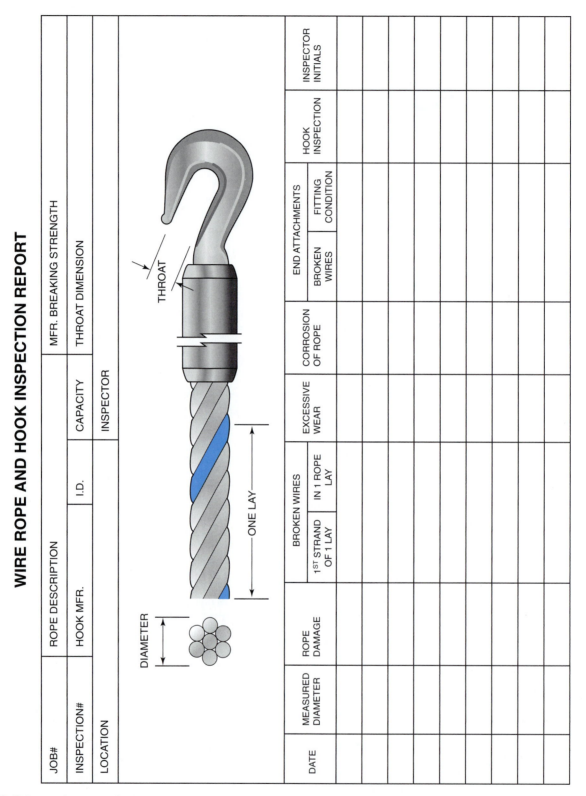

Figure 16–5 Inspection report form.

Figure 16–6 Wire broken by pulling apart. *Courtesy of Wire Rope Corporation of America, Inc.*

Figure 16–7 Wire broken by fatigue. *Courtesy of Wire Rope Corporation of America, Inc.*

Figure 16–8 Results of heavy loads with tight bends. *Courtesy of Wire Rope Corporation of America, Inc.*

Figure 16–9 Strand nicking. *Courtesy of Wire Rope Corporation of America, Inc.*

Figure 16–10 Birdcage caused by sudden release of tension. *Courtesy of Wire Rope Corporation of America, Inc.*

Figure 16–7 shows a wire with a fatigue break. It is recognized by the square end perpendicular to the wire. A fatigue break occurs when the wire is "worked" or flexed back and forth until it becomes hardened and brittle. You can see this kind of fatigue by bending a wire coat hanger back and forth at one point until it breaks.

Figure 16–8 is an example of fatigue failure of a wire rope subjected to heavy loads over small sheaves. The breaks in the valleys of the strands are caused by strand nicking. There may be crown breaks, too.

Figure 16–9 shows a single strand removed from a wire rope subjected to strand nicking. This condition is a result of adjacent strands rubbing against one another. While this is normal in a rope's operation, the nicking can be made worse by high loads, small sheaves, or loss of core support. The ultimate result will be individual wire breaks in the valleys of the strands.

A "birdcage" (Fig. 16–10) is caused by a sudden release of tension and the resulting rebound of rope. These strands and wires will not be returned to their original positions. The rope should be replaced immediately.

Figure 16–11 High strand from wear of surrounding strands. *Courtesy of Wire Rope Corporation of America, Inc.*

Figure 16–12 Kink. *Courtesy of Wire Rope Corporation of America, Inc.*

Figure 16–11 shows a wire rope with a high strand—a condition in which one or more strands are worn before adjoining strands. This is caused by improper socketing or seizing, kinks, or doglegs. It recurs every sixth strand in a six-strand rope.

A kinked wire rope is shown in Figure 16–12. It is caused by pulling down a loop in a slack line during handling, installation, or operation. Note the distortion of the strands and individual wires. This rope must be replaced.

Web Slings

Synthetic web slings (Fig. 16–13) are frequently used because they offer several advantages over wire rope. They are lighter weight and more flexible, and so they are easier to handle. They conform to the shape of the load, and so they are less apt to damage the load and they are less apt to slip. They are not generally affected by water. In addition, they do not corrode, and so they do not stain the surface of the load. They are, however, damaged by exposure to temperatures over 180°F, and they are susceptible to cuts and abrasion. Many synthetic web slings deteriorate if left exposed to sunlight for long periods of time. Just as for wire rope, it is important to know the characteristics of web slings and to inspect them frequently.

Construction

Web slings are made of nylon or polyester. Both make excellent, flexible slings. Nylon slings are damaged by acids but resist caustics. Polyester slings are damaged by caustics but resist acids. Both materials are used for all the common configurations of slings, and the inspection criteria for both are the same. Figure 16–14 shows the common types of synthetic slings. Some web slings have eyes sewn into the ends, and some have hardware fittings on the ends. Web slings with hardware fittings usually have a male end and a female end. The male end can be passed through an opening in the female

Figure 16–13 Two synthetic slings being used with a chain. *Courtesy of Lift-All Company, Inc.*

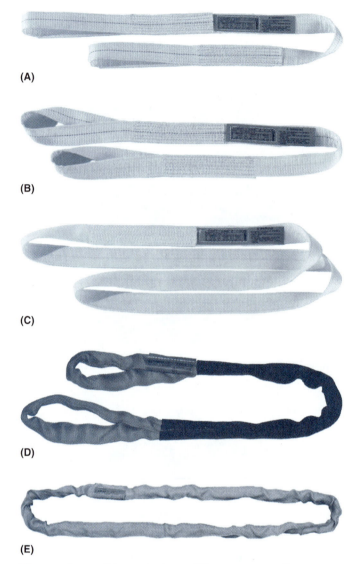

Figure 16–14 **(A)** Standard eye, **(B)** twisted eye, **(C)** endless, **(D)** eye-and-eye round sling, and **(E)** endless round sling. *Courtesy of Lift-All Company, Inc.*

end for a choker hitch. A round sling is made of nylon or polyester, but—as its name says—it is round, not flat. All synthetic slings have warning markers embedded in the material (Fig. 16–15). When the sling wears to the point where the marker is exposed, the sling must be taken out of service.

Inspection of Synthetic Slings

OSHA regulations state that "each day before being used, the sling and all fastenings and attachments shall be inspected for damage or defects by a competent person designated by the employer. Additional inspections shall be performed during sling use, where service conditions warrant." In other words, you should visually inspect your sling before each lift. Figure 16–16 shows defects that might appear in web slings. If you see damage, take the sling out of service and do not use it until it has been approved to be put back in service by a competent person.

Here are reasons for removing a sling from service according to the American National Standards Institute (ANSI B30.9):

- Acid or caustic burns
- Melting or charring of any part of the sling

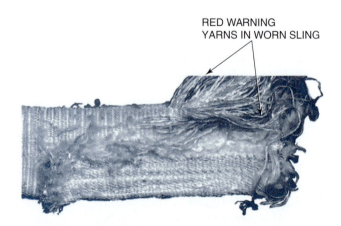

RED WARNING YARNS IN WORN SLING

Figure 16–15 **Red warning yarns.** *Courtesy of Lift-All Company, Inc.*

Web Slings

INSPECTION CRITERIA

⚠ **WARNING** Read Definition on page 3

Inspection Criteria for Synthetic Web Slings
Refer to illustrations of damaged webbing

Remove from service if any of the following is visible:

- Capacity tag is missing or illegible

- Red core warning yarns are visible

- Sling shows signs of melting, charring or chemical damage

- End fittings are excessively pitted, corroded, distorted, cracked or broken

- Cuts on the face or edge of webbing

- Holes, tears, snags or crushed web

- Signs of excessive abrasive wear

- Broken or worn threads in the stitch patterns

- Any other visible damage which causes doubt as to its strength

Red Core Yarns - warn of dangerous sling damage. All *Lift-All* Web Slings shown in this section of the catalog have this warning feature. When red yarns are visible, the sling should be removed from service immediately. The red core yarns become exposed when the sling surface is cut or worn through the woven face yarns. For other inspection criteria see OSHA/Manufacturer regulations on pages 6 through 11.

Examples of Web Sling Abuse

Most of the damage shown here would cause immediate catastrophic failure of the sling. Not all of the damage you will see will be this obvious or extreme, but still requires removal from use.

Elasticity - The stretch characteristics of web slings depends on the type of yarn and the web finish. Approximate stretch at RATED SLING CAPACITY is:

NYLON		POLYESTER	
Treated	10%	Treated	7%
Untreated	6%	Untreated	3%

Prior to sling selection and use, review and understand the "Help" section.

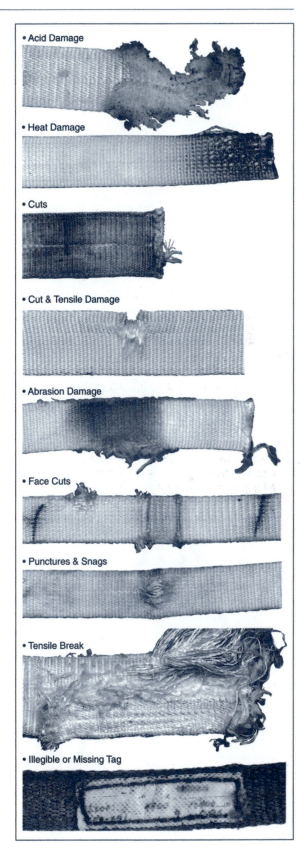

- Acid Damage
- Heat Damage
- Cuts
- Cut & Tensile Damage
- Abrasion Damage
- Face Cuts
- Punctures & Snags
- Tensile Break
- Illegible or Missing Tag

Figure 16–16 **Defects in synthetic slings.** *Courtesy of Lift-All Company, Inc.*

- Holes, tears, cuts, or snags
- Broken or worn stitching in load-bearing splices
- Excessive abrasive wear
- Knots in any part of the sling
- Excessive pitting or corrosion, or cracked, distorted, or broken fittings
- Other visible damage that causes doubt as to the strength of the sling

In addition, slings should not be used if you see any of the following:

- Red warning markers are showing.
- The sling is distorted in any way.
- The identification tag is not readable.
- The sling is loaded beyond its rated capacity for whatever reason.

While most of these standards are very specific regarding reasons for removal, others require your good judgment. The critical areas to watch are wear to the sling body, the wear or unraveling of the selvage edge of webbing, and the condition of the sling eyes.

Chain Slings

Chain slings (Fig. 16–17) are used when there are especially heavy loads or when the sling will be subjected to heat that would damage the core of a wire rope or the synthetic material in a web sling. Chain slings are used less often on most construction jobs, because they are heavy and the chain links may damage the load.

OSHA requires that all chain slings be inspected by a trained competent person at least once a year, but anyone who is rigging loads with a chain sling should be able to recognize the damage shown in Figure 16–18. If any of these signs of damage are present, the sling should not be used. Figures 16–19 through 16–35 show a list of do's and don'ts for the use of slings.

General Requirements for Use of All Slings

The following are OSHA and manufacturer requirements that apply to all slings.

Effect of Angle on Sling Capacity

Using slings at an angle can become deadly if the angle is not taken into consideration. Consider the two slings shown in Figures 16–36 and 16–37 (page 197). The straight vertical hitch, with a load of 1,000 pounds, puts 1,000 pounds of stress on the sling. The double choker hitch is also lifting a 1,000-pound load. Not only is that 500 pounds of downward force on each leg, but, in addition, there is a strong force pulling the ends of the sling toward the middle. Thus, the actual stress on each leg of the sling is much more than 500 pounds.

Follow these steps to calculate the capacity sling needed:

1. Determine the weight that the sling will be lifting (LW).
2. Calculate the tension factor (TF) by dividing the sling height (H) by the sling length (L).
3. Multiply the lifting weight (LW) by the tension factor (TF) to find the minimum sling rating needed for that lift.

Figure 16–17 **Chain sling.** *Courtesy of Lift-All Company, Inc.*

BENT LINKS

USUALLY CAUSED BY BENDING OVER SHARP EDGES OF A LOAD.

Figure 16–18 **Chain defects.** *Courtesy of Lift-All Company, Inc.*

GOUGED LINKS

DAMAGED BY A HEAVY OBJECT BEING DRAGGED OVER OR DROPPED ON THE CHAIN.

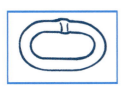

STRETCHED LINKS

INDICATES THE SLING HAS BEEN EXTREMELY OVERLOADED OR SUBJECTED TO SHOCK LOADING. THESE LINKS WOULD NOT HINGE FREELY WITH ADJACENT LINKS.

WORN LINKS

EXCESSIVE WEAR, ESPECIALLY AT THE BEARING POINTS, SERIOUSLY WEAKENS THE CHAIN.

MULTIPLE TYPES OF DAMAGE

WORN, GOUGED, AND BENT LINKS

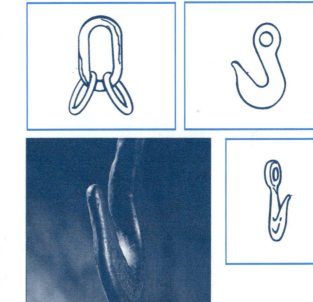

DAMAGED HOOKS AND ATTACHMENTS

THIS HOOK TIP HAS BEEN BENT ONE DIRECTION AND THE EYE ANOTHER. THE TIP WAS PROBABLY POINT LOADED AND THE EYE BENT BY BEING RIGGED OVER THE EDGE OF A LOAD.

Figure 16–18 Continued. *Courtesy of Lift-All Company, Inc.*

EXAMPLE **1**

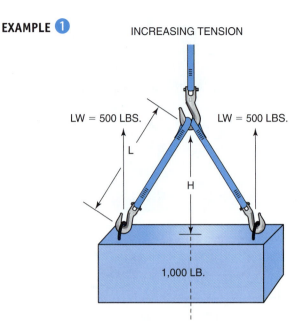

INCREASING TENSION

LW = 500 LBS.   LW = 500 LBS.

L

H

1,000 LB.

EXAMPLE:
LOAD WEIGHT = 1,000 LBS.
RIGGING - 2 SLINGS IN VERTICAL HITCH
LIFTING WEIGHT (LW) PER SLING = 500 LB.
MEASURED LENGTH (L) = 10 FT.
MEASURED HEIGHT (H) = 5 FT.
TENSION FACTOR (TF) = 10 (L) ÷ 5 (H) = 2.0
MINIMUM VERTICAL RATED CAPACITY REQUIRED
FOR THIS LIFT = 500 (LW) × 2.0 (TF) = 1,000 LB.
PER SLING

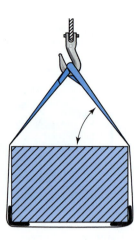

Figure 16–21 Angle of lift must be considered in all lifts. See Effect of Angle on Sling Capacity on page 192. *Courtesy of Lift-All Company, Inc.*

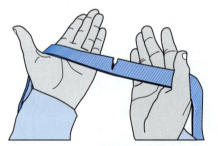

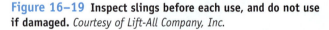

Figure 16–19 Inspect slings before each use, and do not use if damaged. *Courtesy of Lift-All Company, Inc.*

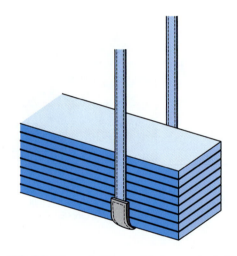

Figure 16–22 Slings shall be padded or protected from sharp edges on their loads. *Courtesy of Lift-All Company, Inc.*

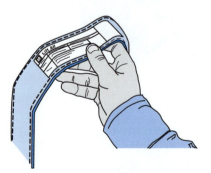

Figure 16–20 Slings shall not be loaded in excess of their rated capacities. Rated capacities (working load limits) must be shown by markings or tags attached to all slings. *Courtesy of Lift-All Company, Inc.*

Figure 16–23 Loads must be rigged to prevent slippage. *Courtesy of Lift-All Company, Inc.*

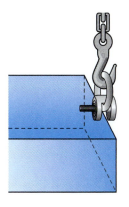

Figure 16–24 **Slings shall be securely attached to their loads.** *Courtesy of Lift-All Company, Inc.*

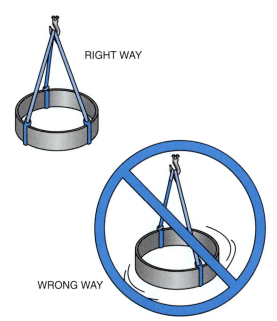

RIGHT WAY

WRONG WAY

Figure 16–25 **Lift must be stable with respect to center of gravity.** *Courtesy of Lift-All Company, Inc.*

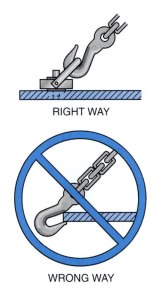

RIGHT WAY

WRONG WAY

Figure 16–26 **Do not point-load hooks. Center load in base of hook.** *Courtesy of Lift-All Company, Inc.*

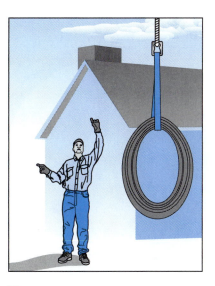

Figure 16–27 **Suspended loads shall be kept clear of all obstructions, and all persons shall be kept clear of loads to be lifted and suspended load.** *Courtesy of Lift-All Company, Inc.*

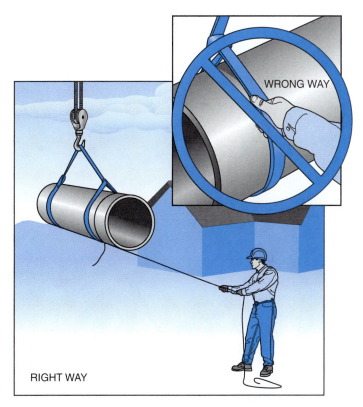

WRONG WAY

RIGHT WAY

Figure 16–28 **Hands and fingers shall not be placed between the sling and the load while the sling is being tightened around the load. After lifting, the load should not be pushed or guided by employees' hands directly on the load. Ropes or tag lines should be attached for this purpose.** *Courtesy of Lift-All Company, Inc.*

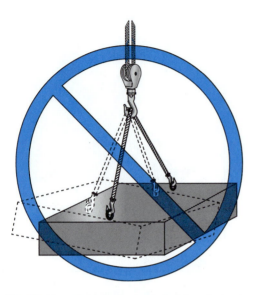

Figure 16-29 **Do not shock the load. Jerking the load could overload the sling and cause it to fail.** *Courtesy of Lift-All Company, Inc.*

Figure 16-31 **Temperature and chemical environment must be considered.** *Courtesy of Lift-All Company, Inc.*

RIGHT WAY

WRONG WAY

Figure 16-30 **A sling shall not be pulled from under a load when the load is resting on the sling. Before a load is lifted, a place should be prepared where it is to be put down. Lumber can be used to allow a space to remove the sling and prevent shifting of the load.** *Courtesy of Lift-All Company, Inc.*

Figure 16-32 **Slings shall not be shortened with knots, bolts, or makeshift devices.** *Courtesy of Lift-All Company, Inc.*

Figure 16-33 **Sling legs shall not be kinked or twisted.** *Courtesy of Lift-All Company, Inc.*

Figure 16–34 Slings shall not be dragged on the floor or ground. *Courtesy of Lift-All Company, Inc.*

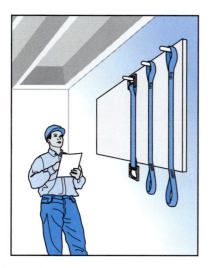

Figure 16–35 Slings shall be stored in cool, dark, dry areas, preferably on racks. *Courtesy of Lift-All Company, Inc.*

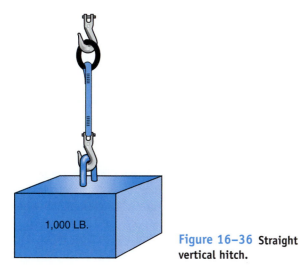

Figure 16–36 Straight vertical hitch.

Figure 16–37 Double choker hitch.

Hardware Attachments

Hooks

Hooks are permanently attached to slings and are used to attach the sling to the load. There are several types of hooks, and each has a specific purpose (Fig. 16–38). Most hooks are eye hooks. That is, they have an eye for attachment to the sling or hoist. A reversed-eye hook has an eye at a right angle to the plane of the hook. A grab hook has a narrow throat so that it will grab a link in the chain and will not slip. A foundry hook has a greater throat opening so that the chain links or wire rope will slide through the hook. Many hooks used for slings are latch hooks. Latch hooks have a latch that helps prevent the load from slipping out of the hook. A clevis hook has a clevis-style pin instead of an eye for attachment to the hoist or sling. A choker hook has a barrel-like opening through which a wire rope can slide when used in a choker hitch.

Other Sling Hardware

A shackle (Fig. 16–39) is used to attach a sling to a load or to join slings together. There are several types of shackles: chain shackles, anchor shackles, web shackles, and specialty shackles. Shackles should be inspected before each use just like other sling components.

Figure 16–38 **Several types of hooks: (A) standard eye hook, (B) grab hook, (C) latch hook, and (D) choker hook.** *Courtesy of Lift-All Company, Inc.*

Figure 16–39 **Clevis-type shackle.** *Courtesy of Lift-All Company, Inc.*

PLAIN   SHOULDERED

Figure 16–40 **Lifting eye.**

- stress cracks.
- rust or corrosion.
- damaged threads.

Thimbles

A **thimble** is used inside the eye of a wire rope sling to decrease wear and to protect the wire rope from kinking (Fig. 16–41). The thimble is inspected for wear, rust, or other visible damage each time the sling is inspected.

Wire Rope Clips

Wire rope clips are used to fasten a loop in a wire rope (Fig. 16–42). Clips come in different sizes to match the size of the rope. It is important to use a torque wrench when tightening rope clips so that they will be tight enough to provide maximum holding power, but not so tight as to damage the rope.

Lifting eyes (Fig. 16–40) can be straight eye bolts or swivel eyes. Straight eye bolts should only be used where the load will pull straight along the axis of the eye bolt. Where the load must be applied at an angle, a swivel eye should be used. Lifting eyes should be inspected for:

- wear that has removed part of the material.
- bent shank or eye.

Figure 16–41 **Thimble.** *Courtesy of Lift-All Company, Inc.*

Figure 16–42 **Wire rope clips.**

Figure 16–43 **Tag line.**

Crane Operations

When lifting with a crane, it is important to ensure the safety of all people on the site and to protect the property by following a few safety rules. One person on the ground should be designated to direct the operations. That person is responsible for everyone else's safety. Anyone who is not directly involved in the operation should be cleared from the path of the load and cleared from the area of the crane. Only the designated person should give signals to the crane operator. When a long load might swing or be blown by the wind, use a tag line to control it (Fig. 16–43). A **tag line** is a rope tied to the load and used to guide the load from the ground.

The only person in the landing area should be the person controlling the tag line to position the load as it comes to rest.

CAUTION

CAUTION: Never attempt to hold a load or position it with your hands. The area where the load will be placed should be prepared before the load is lifted. If blocking is needed to support the load, it should be prepared and placed before the lift is begun.

CAUTION

CAUTION: Attempting to move the load into place by hand can cause serious back injury.

The crane operator receives instructions from the rigger either by radio or by hand signals. The hand signals shown in Fig. 16–44 have been approved by the American National Standards Institute.

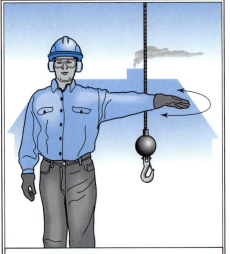

STOP. ARM EXTENDED, PALM DOWN, MOVE HAND RIGHT AND LEFT.

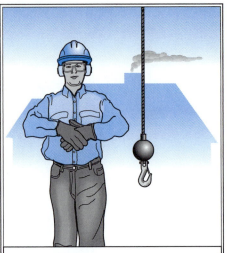

DOG EVERYTHING. CLASP HANDS IN FRONT OF BODY.

MOVE SLOWLY. USE ONE HAND TO GIVE ANY MOTION SIGNAL AND PLACE OTHER HAND MOTIONLESS IN FRONT OF HAND GIVING THE MOTION SIGNAL, (HOIST SLOWLY SHOWN AS EXAMPLE).

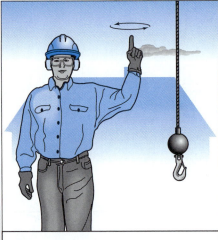

HOIST. WITH FOREARM VERTICAL, FINGER POINTING UP, MOVE HAND IN SMALL HORIZONTAL CIRCLES.

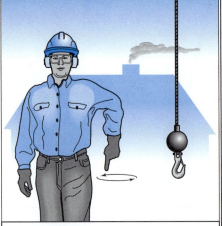

LOWER. WITH ARM EXTENDED DOWNWARD, FOREFINGER POINTING DOWN, MOVE HAND IN SMALL HORIZONTAL CIRCLES

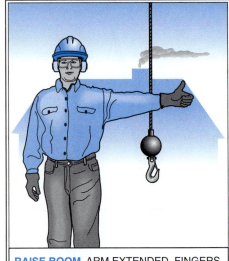

RAISE BOOM. ARM EXTENDED, FINGERS CLOSED THUMB POINTING UPWARD.

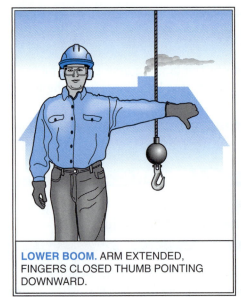

LOWER BOOM. ARM EXTENDED, FINGERS CLOSED THUMB POINTING DOWNWARD.

SWING. ARM EXTENDED, POINT WITH FINGER IN DIRECTION OF SWING OF BOOM

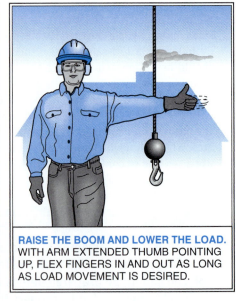

RAISE THE BOOM AND LOWER THE LOAD. WITH ARM EXTENDED THUMB POINTING UP, FLEX FINGERS IN AND OUT AS LONG AS LOAD MOVEMENT IS DESIRED.

Figure 16–44 Hand signals for controlling crane operations.

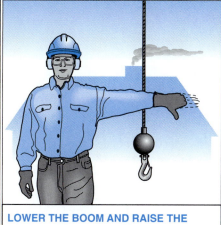

LOWER THE BOOM AND RAISE THE LOAD. WITH ARM EXTENDED, THUMB POINTING DOWN, FLEX FINGERS IN AND OUT AS LONG AS LOAD MOVEMENT IS DESIRED.

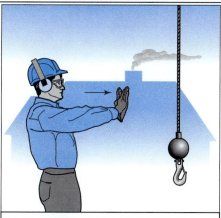

TRAVEL. (RAIL MOUNT OR TROLLEY.) ARM EXTENDED FORWARD. HAND OPEN AND SLIGHTLY RAISED, MAKING PUSHING MOTION IN DIRECTION OF TRAVEL.

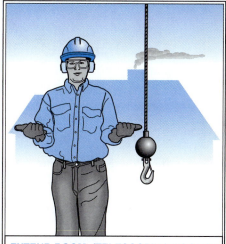

EXTEND BOOM. (TELESCOPING BOOMS.) BOTH FISTS IN FRONT OF BODY WITH THUMBS POINTING OUTWARD.

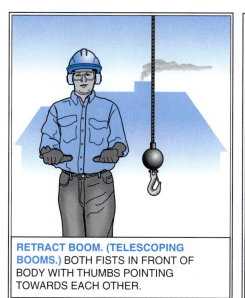

RETRACT BOOM. (TELESCOPING BOOMS.) BOTH FISTS IN FRONT OF BODY WITH THUMBS POINTING TOWARDS EACH OTHER.

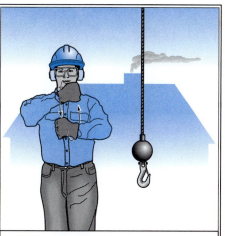

TRAVEL. ARM EXTENDED FORWARD, HAND OPEN AND SLIGHTLY RAISED, MAKING RUNNING MOTION IN DIRECTION OF TRAVEL.

TRAVEL. (ONE TRACK.) LOCK THE TRACK ON SIDE INDICATED BY RAISED FIST. TRAVEL OPPOSITE TRACK IN DIRECTION INDICATED BY CIRCULAR MOTION ON OTHER FIST, ROTATED VERTICALLY IN FRONT OF BODY. (FOR CRAWLER CRANES ONLY.)

Figure 16–44 Continued.

Review Questions

1. Where can you find the type of material of which a synthetic sling is made?

2. Where can you find the rated capacity of an alloy steel chain sling?

3. What is the purpose of a wear pad on a wire rope sling?

4. How is the word "lay" used as measurement in wire rope?

5. What causes a birdcage, and what should be done about it?

6. List three defects that would be reasons to remove a wire rope from service.

7. What is one advantage of web slings over wire rope slings?

8. Where can you find the name of the manufacturer of a web sling?

9. What is indicated by red yarns showing on the surface of a web sling?

10. List three defects that would be reasons to remove a web sling from service.

11. Who should inspect synthetic slings, and how often?

12. List three defects that would be reasons to remove a steel alloy chain sling from service.

13. Where can you find the rated capacity of a steel alloy chain sling?

14. Where is a thimble used in rigging?

15. What rated capacity is needed for a sling to lift the load shown in Figure 16–45?

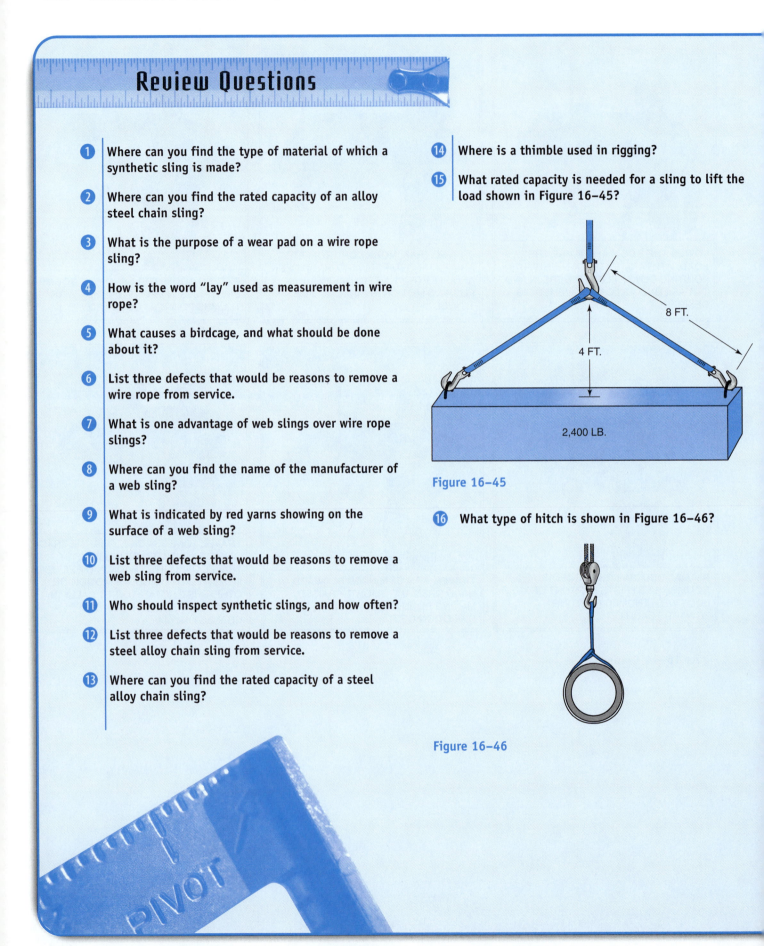

8 FT.

4 FT.

2,400 LB.

Figure 16–45

16. What type of hitch is shown in Figure 16–46?

Figure 16–46

17 What type of hitch is shown in Figure 16–47?

18 What is indicated by the hand signal in Figure 16–48?

19 What is indicated by the hand signal in Figure 16–49?

20 Who is responsible for the safety of all the workers in the area of a crane operation?

Figure 16–48

Figure 16–47

Figure 16–49

Activities

CRANE OPERATIONS

Ideally, you should gain experience directing crane operations with a real load and a real crane, but that is not realistic in the classroom. The classroom crane used for this activity will give you an opportunity to use ANSI-approved hand signals and to simulate some of the conditions and concerns you would encounter with a real crane. The classroom crane moves in all the same ways as a real hydraulic boom crane. It just lacks the size, strength, and automatic features of a real one.

You will need to build the classroom crane shown in Figure 16–50 if your classroom does not already have a crane.

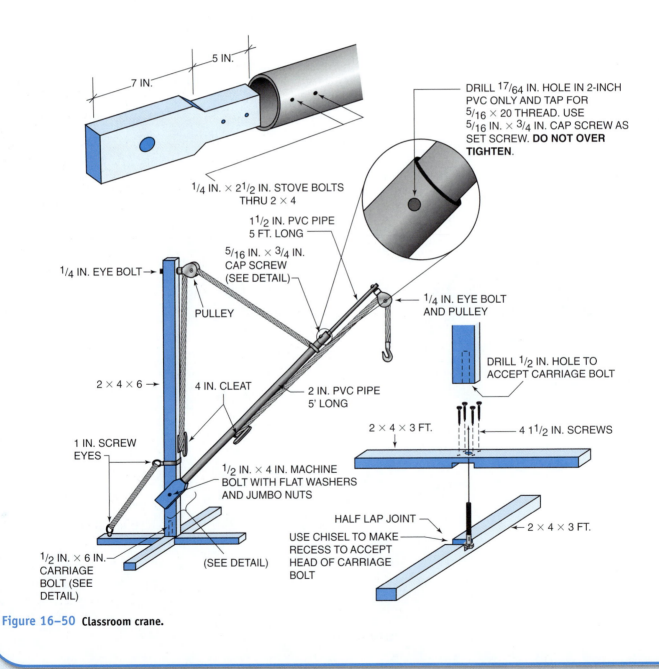

Figure 16–50 Classroom crane.

Materials for Classroom Crane

- Three pieces of lumber 2×4×8 ft
- 2-inch PVC pipe, 5 feet long
- 1½-inch PVC pipe 5 feet long
- Two small pulleys, suitable for 1/4-inch rope
- 50 feet of 1/4-inch rope or sash cord
- Two 4-inch rope cleats
- 1/4-inch × 4-inch-long eye bolt with nut and washer
- 1/4-inch × 2-inch-long eye bolt with nut
- 5/16-inch × 3/4-inch UNC cap screw
- 1/4-inch × 2½-inch stove bolts with nuts
- Two 1-inch screw eyes
- 1/2-inch × 6-inch carriage bolt
- 1/2-inch × 4-inch machine bolt with two flat washers and two nuts

Sling Materials

Slings suitable for use with the classroom crane can be made from 1-inch-wide nylon webbing or 3/8-inch rope. Eyes in nylon web slings can be made by riveting if sewing equipment is not available. Wire rope clips can be used on 3/8-inch rope.

PROCEDURE

This activity will require a three-person team: a crane operator, a rigger who will give hand signals, and a tender for the tag line.

1. Construct the crane as shown in Figure 16–50, or use an existing crane. The set screw at the outer end of the 2-inch pipe should only be tightened enough to hold the 1½-inch pipe in position—hand tight is probably enough, but no more than 1/8 turn with a wrench after hand tightening or else the threads may strip. This is the telescoping boom found on many portable cranes. Use sandbags on the base to prevent the crane from tipping over. The farther you extend the boom, the more weight you will need on the base. Do not attempt to lift more than 30 pounds with the classroom crane.
2. Position the crane in an area where you can lift and move your load without endangering other people or expensive equipment and furnishings in the room.

3. Select a sling that is suitable for the load you will be lifting. You will be lifting an 8-foot-long 2×6, and so the load will not be great, but you will want to be able to balance your load.
4. Assume your load is a heavy beam that has just been delivered on a truck. It rests on blocks at each end, and so you can pass the sling underneath. Prepare a landing area, with blocks on which to set the beam. The beam should be within 4 feet of the crane to start. You will lift it over a barrier (a workbench, chair, or any other artificial barrier will do) and land it 7 or 8 feet away from the crane and to one side (Fig. 16–51).
5. Attach a suitable sling and a tag line.
6. Check the path of the load to ensure that it is clear and that no one will be in danger. Check to see that the crane boom can be extended and swung over the landing area without any obstructions.
7. Following only hand signals from the rigger, slowly lift the load about 3 feet off the floor and maneuver the crane so that the load can be landed on the prepared blocks. The person with the tag line will have to turn the load as necessary to line it up with the blocks.

CAUTION: Never place your hands on a load while it is being lifted.

Figure 16–51 **Placements for crane activity.**

Print Reading

Success Stories

Chris Ziemann

TITLE

Co-owner of BZ Construction and Design, a residential and light commercial construction company.

EDUCATION

Chris graduated from Southeast Community College, Milford Campus, with an associate's degree in building construction technology.

HISTORY

Chris's interest in construction began when he was young and worked with his dad on days when he was not in school. He took Computer-Aided Drafting (CAD) and shop classes in high school. He proved his talent by being student of the quarter at tech school and by graduating with distinction. After completing his education, Chris began working with BZ Construction and Design and later became co-owner of the business.

ON THE JOB

"I enjoy the opportunity of working with homeowners. I work 7:30 a.m. to 5 p.m. on all indoor and outdoor projects. As co-owner, my job is to complete the project in a timely manner and to make sure the customers are completely satisfied with the work we have done." Chris's responsibilities include making construction drawings, building the project as the drawings indicate, shingling, trimming, and building decks. As a carpenter, he interacts with all subcontractors on a daily basis in order to keep projects on track.

BEST ASPECTS

"I enjoy seeing how everything comes together on a project," says Chris. "It proves that when people work together a lot can be accomplished. The beginning of a project takes a lot of time and effort, but when the project is complete, it is something you can be proud of and say you were a part of."

CHALLENGES

Chris is challenged by working with customers on a daily basis and trying to understand what they want their final projects to look like. "I'm constantly continuing my education by learning new things from other people's ideas as well as by forming my own ideas," says Chris.

IMPORTANCE OF EDUCATION

Without the education that he received, Chris feels he wouldn't have the communication skills or the work ethic that he has today. He works every day with numbers and calculations—which makes some days very challenging. "I have a lot of respect and appreciation for my SCC instructors. They taught me how to read and make drawings, which has enabled me to get where I am today. If you're not learning, you're not living."

FUTURE OPPORTUNITIES

Chris anticipates starting his own drafting business by taking what he has learned through schooling and on-the-job training and putting it to use. He also hopes to continue working with all of his SCC instructors, and to one day become an instructor himself.

WORDS OF ADVICE

"What is memorized today is forgotten tomorrow. What is understood today is with you for a lifetime."

Chapter 17 Views

210

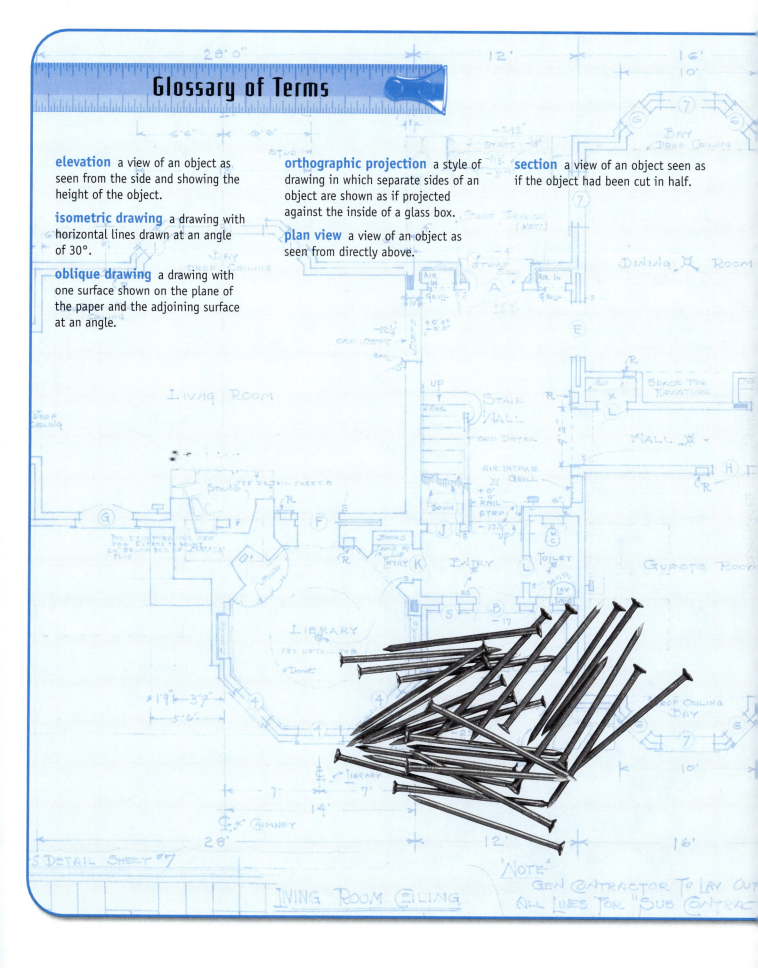

Glossary of Terms

elevation a view of an object as seen from the side and showing the height of the object.

isometric drawing a drawing with horizontal lines drawn at an angle of 30°.

oblique drawing a drawing with one surface shown on the plane of the paper and the adjoining surface at an angle.

orthographic projection a style of drawing in which separate sides of an object are shown as if projected against the inside of a glass box.

plan view a view of an object as seen from directly above.

section a view of an object seen as if the object had been cut in half.

Buildings and their parts can be shown in drawings in a variety of different views. An isometric drawing shows three sides of the object in a single drawing. Another type of drawing, called an orthographic drawing or orthographic projection, shows the sides of the object as though they are viewed straight on, so that only one surface is seen in a single drawing. No matter what your involvement with the home building industry, you will probably encounter both types of views in your work.

Isometric Drawings

A useful type of pictorial drawing for construction purposes is the **isometric drawing.** In an isometric drawing, vertical lines are drawn vertically and horizontal lines are drawn at an angle of 30° from horizontal (Fig. 17–1). All lines on one of these isometric axes are drawn in proportion to their actual length. Isometric drawings tend to look out of proportion because we are used to seeing the object appear smaller as it gets farther away.

Isometric drawings are often used to show construction details (Fig. 17–2). The ability to draw simple isometric sketches is a useful skill for communicating on the job site. Try sketching a brick in isometric as shown in Figure 17–3.

1 Sketch a Y with the top lines about 30° from horizontal.

2 Sketch the bottom edges parallel to the top edges.

3 Mark off the width on the left top and bottom edges. This will be about twice the height.

4 Mark off the length on the right top and bottom edges. The length will be about twice the width.

5 Sketch the two remaining vertical lines and the back edges.

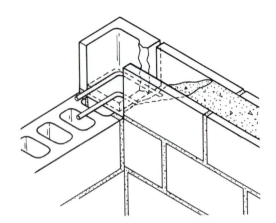

Figure 17–2 **Isometric construction detail.** *Courtesy of W.D. Farmer.*

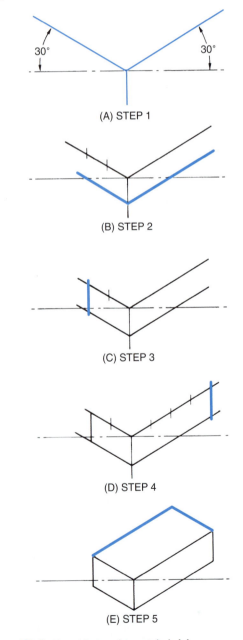

(A) STEP 1

(B) STEP 2

(C) STEP 3

(D) STEP 4

(E) STEP 5

Figure 17–3 **Sketching an isometric brick.**

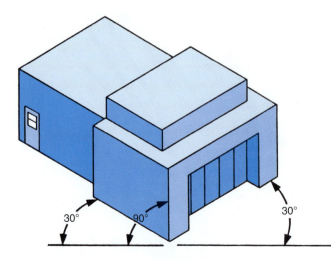

Figure 17–1 **Isometric of a building.**

Other isometric shapes can be sketched by adding to or subtracting from this basic isometric brick (Fig. 17–4). Angled surfaces are sketched by locating their edges and then connecting them.

Oblique Drawings

When an irregular shape is to be shown in a pictorial drawing, an **oblique drawing** may be best. In oblique drawings, the most irregular surface is drawn in proportion as though it were flat against the drawing surface. Parallel lines are added to show the depth of the drawing (Fig. 17–5).

Orthographic Projection

To show all information accurately and to keep all lines and angles in proportion, most construction drawings are drawn by **orthographic projection.** Orthographic projection is often explained by imagining the object to be drawn inside a glass box. The corners and the lines representing the edges of the object are then projected onto the sides of the box (Fig. 17–6). If the box is unfolded, the images projected onto its sides will be on a single plane, as on a sheet of paper (Fig. 17–7). In other

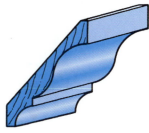

(A) FRONT VIEW OF CROWN MOLDING

(B) OBLIQUE VIEW OF CROWN MOLDING

Figure 17–5 Oblique drawing.

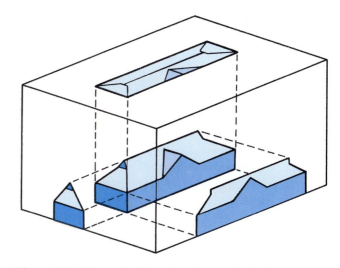

Figure 17–6 House inside a glass box: method of orthographic projection of roof, front, and end.

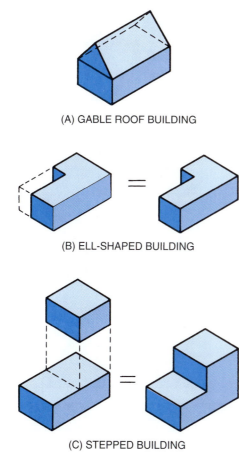

(A) GABLE ROOF BUILDING

(B) ELL-SHAPED BUILDING

(C) STEPPED BUILDING

Figure 17–4 Variations on the isometric brick.

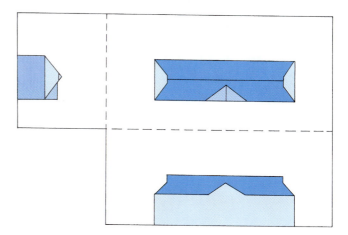

Figure 17–7 The glass box unfolded.

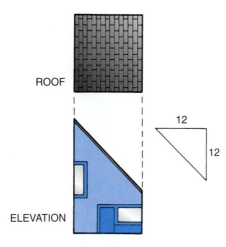

ROOF

ELEVATION

12

12

12

3

Figure 17–8 Views of two shed roofs.

words, in an orthographic projection each view of an object shows only one side (or top or bottom) of the object.

All surfaces that are parallel to the plane of projection (the surface of the box) are shown in proportion to their actual size and shape. However, surfaces that are not parallel to the plane of projection are not shown in proportion. For example, both of the roofs in the top views of Figure 17–8 appear to be the same size and shape, but they are quite different. To find the actual shape of the roof, you must look at the end view.

In construction drawings, the views are called plans and elevations. A **plan view** shows the layout of the object as viewed from above (Fig. 17–9). A set of drawings for a building usually includes plan views of the side (lot), the floor layout, and the foundation. **Elevations** are drawings that show height. For example, a drawing that shows what would be seen standing in front of a house is a building elevation (Fig. 17–10). Elevations are also used to show cabinets and interior features.

Because not all features of construction can be seen in plan views and elevations from the outside of a building, many construction drawings are section views. A section view, usually referred to simply as a **section,** shows what would be exposed if a cut were made through the object (Fig. 17–11). Actually, a floor plan is a type of section view (Fig. 17–12).

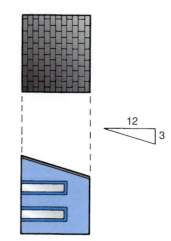

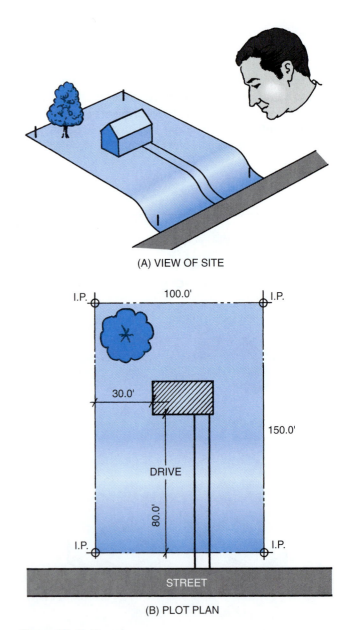

(A) VIEW OF SITE

I.P. — 100.0' — I.P.

30.0'

150.0'

DRIVE

80.0'

I.P. — I.P.

STREET

(B) PLOT PLAN

Figure 17–9 Plan view.

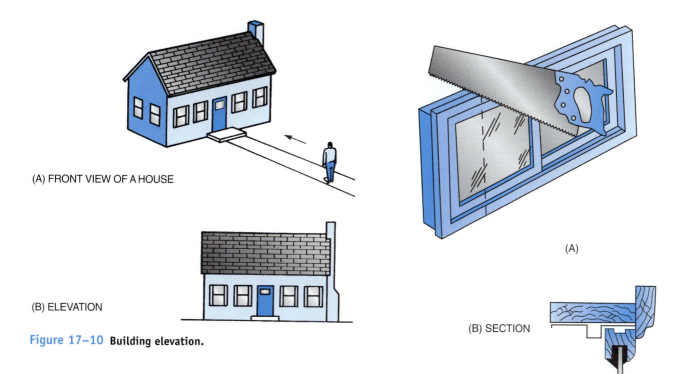

(A) FRONT VIEW OF A HOUSE

(B) ELEVATION

Figure 17–10 **Building elevation.**

(A)

(B) SECTION

Figure 17–11 **Section of a window sash.**

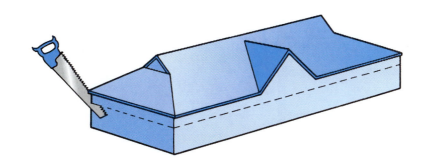

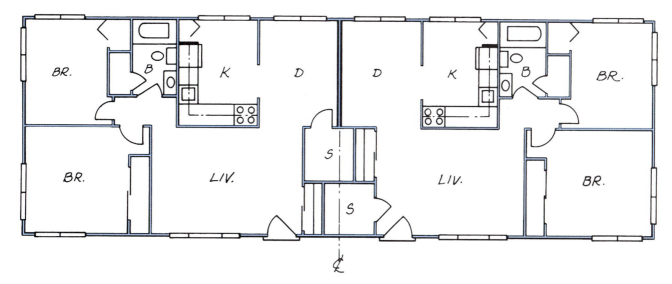

BR.

B

K

D

D

K

B

BR.

BR.

LIV.

S

S

LIV.

BR.

Figure 17–12 **A floor plan is actually a section view of the building.**

Review Questions

1. Identify each of the drawings in Figure 17–13 as oblique, isometric, or orthographic.

2. Identify each of the drawings in Figure 17–14 as an elevation, plan, or section.

3. In the view of the house shown in Figure 17–15, which lines are true length?

4. What type of pictorial drawings are easiest to draw on the job site?

5. What type of drawings are used for working drawings?

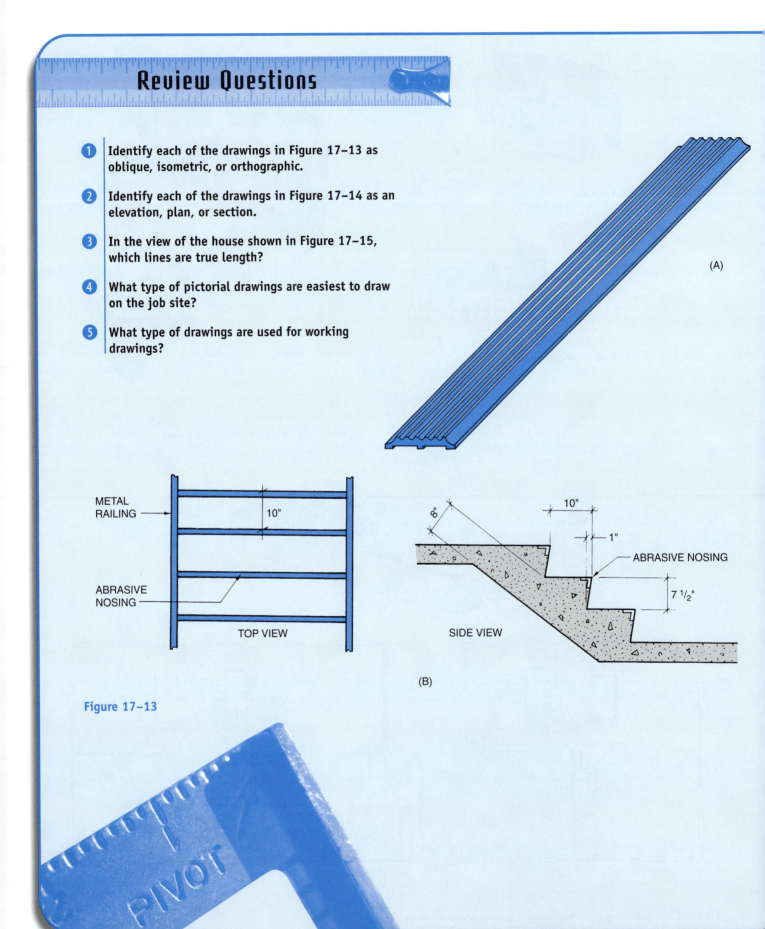

(A)

METAL RAILING → 10"

ABRASIVE NOSING

TOP VIEW

8"

10"

1"

ABRASIVE NOSING

7 1/2"

SIDE VIEW

(B)

Figure 17–13

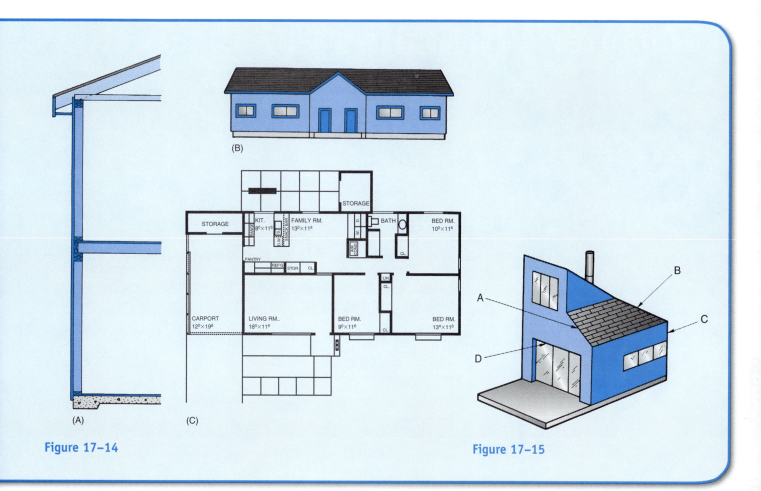

Figure 17–14

Figure 17–15

Activities

As this chapter points out there are many types of drawing views that may be used on construction drawings. To fully understand the drawings, you must know the differences between the types of views. In this activity, you are to review the drawings in Appendix B at the end of the book and answer the following questions:

1. What drawing shows how far the front of the house is from the street? What type of view is that drawing?

2. What type of drawing view shows the layout of the foundation?

3. Explain why the floor plan could be considered a section view.

4. Other than the front, rear, right, and left building elevations, what other drawings use an elevation view?

5. What type of drawing view is used to show details about typical wall construction?

Chapter 18 Scales

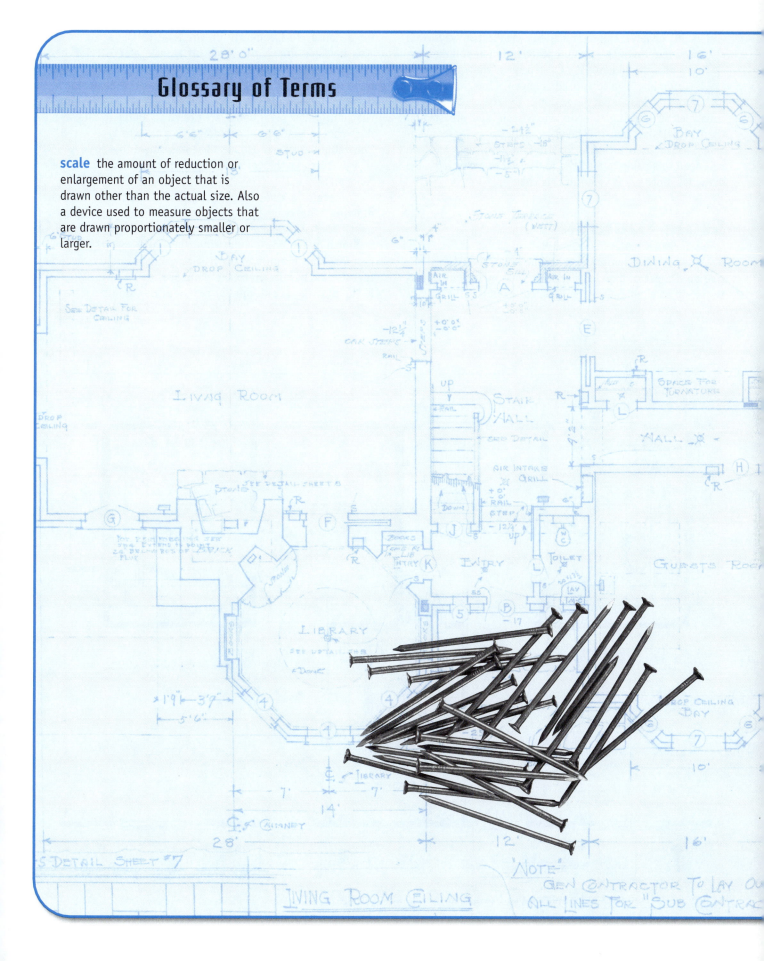

Glossary of Terms

scale the amount of reduction or enlargement of an object that is drawn other than the actual size. Also a device used to measure objects that are drawn proportionately smaller or larger.

n this chapter, the term **scale** is used in two ways. It is used to refer to the amount of reduction or enlargement of a drawing to make it fit the page or show the necessary details. It also is used to mean a tool for measuring drawings that are done to scale. Nearly all construction drawings are drawn to scale and are smaller than actual size. You will need to understand the conventional scales used to make construction drawings, and you will need to be able to read the tools used to measure scaled drawings. Both are covered in this chapter.

Scale Drawings

Because construction projects are too large to be drawn full size on a sheet of paper, everything must be drawn proportionately smaller than it really is. For example, floor plans for a house are frequently drawn 1/48 of the actual size. This is called *drawing to scale*. At a scale of 1/4" = 1'-0", 1/4 inch on the drawing represents 1 foot on the actual building. When it is necessary to fit a large object on a drawing, a small scale is used. Smaller objects and drawings that must show more detail are drawn to a larger scale. The floor plan in Figure 18–1 was drawn to a scale of 1/4" = 1'-0". The detail drawing in Figure 18–2 was drawn to a scale of 3" = 1'-0" to show the construction of one of the walls on the floor plan.

The scale to which a drawing is made is noted on the drawing. The scale is usually indicated alongside or beneath the title of the view. On some drawings, the scale is shown by including a drawing that looks something like a ruler. This graphic scale has graduations representing feet and inches drawn to the scale of the view (Fig. 18–3). If the drawing is enlarged or reduced, the graphic scale is

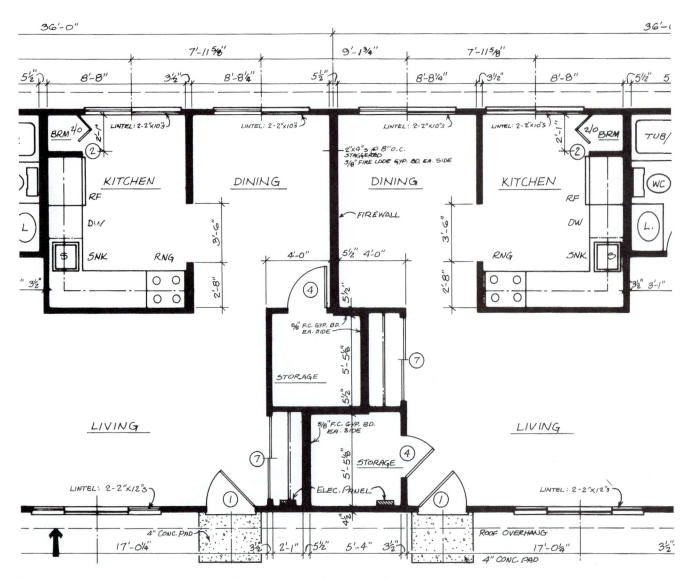

Figure 18–1 Portion of a floor plan with a firewall. *Courtesy of Kurzon Architects.*

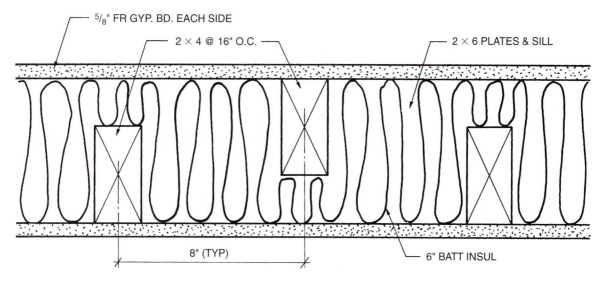

Figure 18–2 **Detail (plan at firewall).**

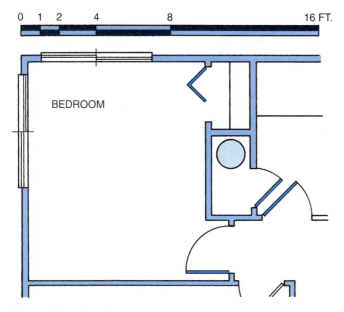

Figure 18–3 **Graphic scale.**

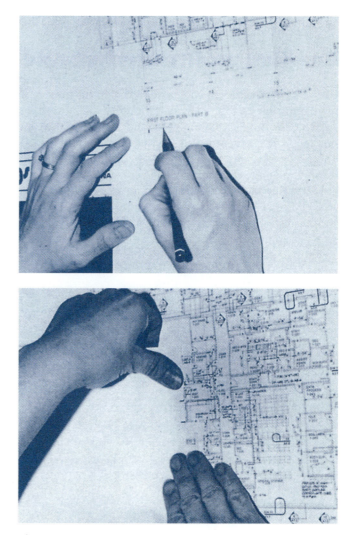

also enlarged or reduced. The graduations on the scale indicator can be marked on the edge of a sheet of paper and then stepped off on the drawing (Fig. 18–4). They may also be transferred to the drawing with dividers (Fig. 18–5).

Figure 18–4 **Marking the graduations on the edge of a piece of paper.**

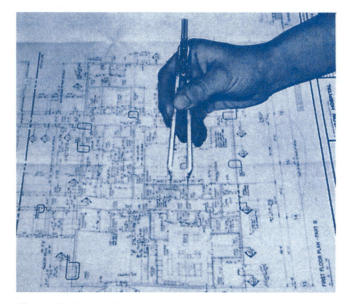

Figure 18–5 Transferring dimensions with dividers.

Reading an Architect's Scale

All necessary dimensions should be shown on the drawings. The instrument used to make drawings to scale is called an *architect's scale* (Fig. 18–6). Measuring a drawing with an architect's or engineer's scale is a poor practice. At small scales it is especially difficult for the drafter to be precise, and any tolerance introduced during drafting can be amplified by trying to use a scale to measure a drawing. The following discussion of how to read an architect's scale is presented to ensure an understanding of the scales used on drawings. The triangular scale includes 11 scales frequently used on drawings.

Full Scale					
3/32"	=	1'-0"	3/16"	=	1'-0"
1/8"	=	1'-0"	1/4"	=	1'-0"
3/8"	=	1'-0"	3/4"	=	1'-0"
1/2"	=	1'-0"	1"	=	1'-0"
1½"	=	1'-0"	3"	=	1'-0"

Two scales are combined on each face, except for the full-size scale, which is fully divided into sixteenths. The combined scales work together because one is twice as large as the other, and their zero points and extra divided units are on opposite ends of the scale.

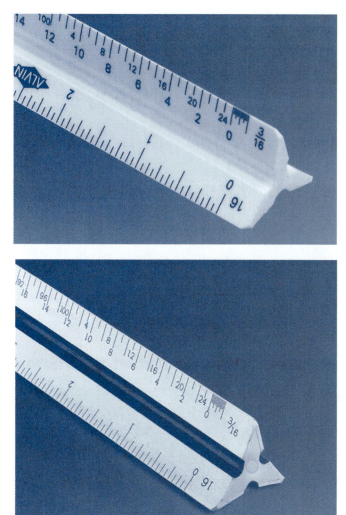

Figure 18–6 Architect's scales. *Courtesy of Alvin & Co., Inc.*

The fraction, or number, near the zero at each end of the scale indicates the unit length in inches that is used on the drawing to represent one foot of the actual building. The extra unit near the zero end of the scale is subdivided into twelfths of a foot (inches) as well as fractions of inches on the larger scales.

To read the architect's scale, turn it to the 1/4-inch scale. The scale is divided on the left from the zero toward the 1/4 mark so that each line represents 1 inch. Counting the marks from the zero toward the 1/4 mark, there are 12 lines marked on the scale. Each one of these lines is 1 inch on the 1/4" = 1'-0" scale.

The fraction 1/8 is on the opposite end of the same scale (Fig. 18–7). This is the 1/8-inch scale and is read in the opposite direction. Notice that the divided unit is only half as

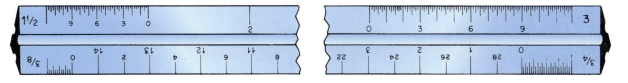

Figure 18–7 **Architect's triangular scale.**

Figure 18–8 **Architect's triangular scale showing 1½-inch and 3-inch scales.**

large as the one on the 1/4-inch end of the scale. Counting the lines from zero toward the 1/8 mark, there are only 6 lines. This means that each line represents 2 inches at the 1/8-inch scale.

Now look at the 1½-inch scale (Fig. 18–8). The divided unit is broken into twelfths of a foot (inches) and also fractional parts of an inch. Reading from the zero toward the number 1½, notice the figures 3, 6, and 9. These figures rep-

resent the measurements of 3 inches, 6 inches, and 9 inches at the 1½" = 1'-0" scale. Between the zero and the three are two long marks representing 1 inch and 2 inches. Between each of these marks are shorter marks dividing the inches into four parts. The spaces between each of the shortest marks represent quarters of an inch. Reading from the zero to the 3, read each line as follows: ¼, ½, ¾, 1, 1¼, 1½, 1¾, 2, 2¼, 2½, 2¾, and 3 inches.

Review Questions

What are the dimensions indicated on the scale in the following figure?

1 _____

2 _____

3 _____

4 _____

5 _____

6 _____

7 _____

8 _____

9 _____

10 _____

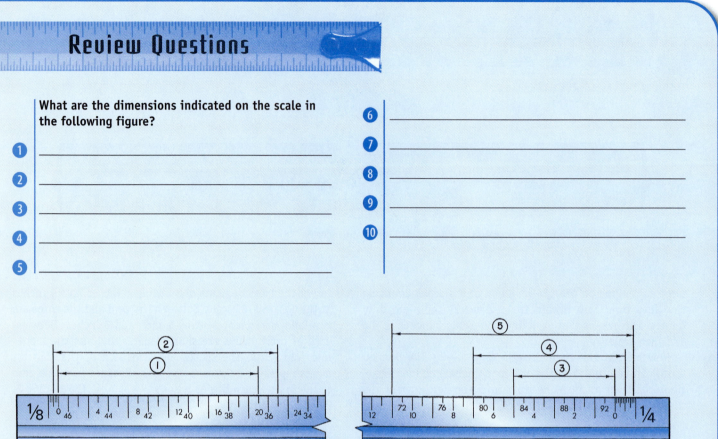

Figure 18–9

Activities

SCALES

What scales are used for the following views of the house in the drawings in Appendix B at the back of the book?

1. Floor plan
2. Plot plan
3. Front elevation
4. Typical wall section

After completing this chapter, the student will be able to identify and understand the meaning of:

- object lines.
- dashed lines (hidden and phantom).
- extension lines and dimension lines.
- centerlines.
- leaders.
- cutting-plane lines.

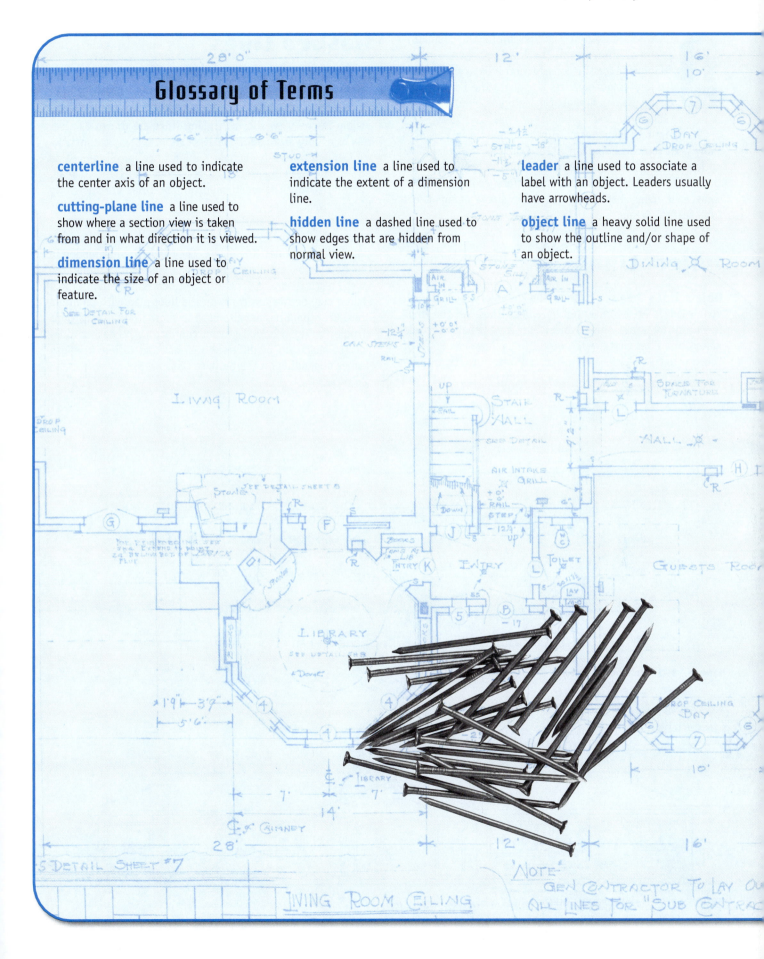

Glossary of Terms

centerline a line used to indicate the center axis of an object.

cutting-plane line a line used to show where a section view is taken from and in what direction it is viewed.

dimension line a line used to indicate the size of an object or feature.

extension line a line used to indicate the extent of a dimension line.

hidden line a dashed line used to show edges that are hidden from normal view.

leader a line used to associate a label with an object. Leaders usually have arrowheads.

object line a heavy solid line used to show the outline and/or shape of an object.

Drawings are used in construction for the communication of information. The drawings, then, serve as a language for the construction industry. The basis for any language is its alphabet. The English language uses an alphabet made up of 26 letters. Construction drawings use an *alphabet of lines* (Fig. 19–1).

The weight or thickness of lines is sometimes varied to show their relative importance. For example, in Figure 19–2 notice that the basic outline of the building is heavier than that of the windows and doors. This difference in line weight sometimes helps distinguish the basic shape of an object from surface details.

Object Lines

Object lines are used to show the shape of an object. All visible edges are represented by object lines. All the lines in Figure 19–2 are object lines. Drawings usually include many solid lines that are not object lines, however. Some of these other solid lines are discussed here. Others are discussed later.

Dashed Lines

Dashed lines have more than one purpose in construction drawings. One type of dashed line, the **hidden line** is used to show the edges of objects that would not otherwise be visible in the view shown. Hidden lines are drawn as a series of evenly sized short dashes (Fig. 19–3). If a construction drawing were to include hidden lines for all concealed edges, the drawing would be cluttered and hard to read. Therefore, only the most important features are shown by hidden lines.

Another type of dashed line is used to show important overhead construction (Fig. 19–4). These dashed lines are called *phantom lines*. The objects they show are not hidden in the view—they are simply not in the view. For example, the most practical way to show exposed beams on a living room ceiling may be to show them on the floor plan with phantom lines. Phantom lines are also used to show alternate positions of objects (Fig. 19–5). To avoid confusion, the dashed lines may be made up of different weights and different-length dashes depending on the purpose (Fig. 19–6).

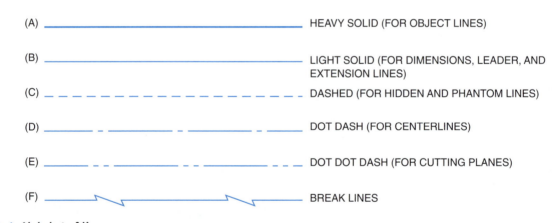

(A) ——————————————————— HEAVY SOLID (FOR OBJECT LINES)

(B) ——————————————————— LIGHT SOLID (FOR DIMENSIONS, LEADER, AND EXTENSION LINES)

(C) – – – – – – – – – – – – – – DASHED (FOR HIDDEN AND PHANTOM LINES)

(D) —— – —— · —— · —— · —— DOT DASH (FOR CENTERLINES)

(E) —— · · —— · · —— · · —— DOT DOT DASH (FOR CUTTING PLANES)

(F) ——⁄\——⁄\—— BREAK LINES

Figure 19–1 **Alphabet of lines.**

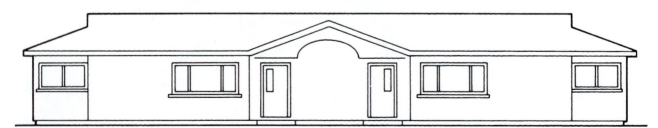

Figure 19–2 **Elevation outlined.** *Courtesy of Kurzon Architects.*

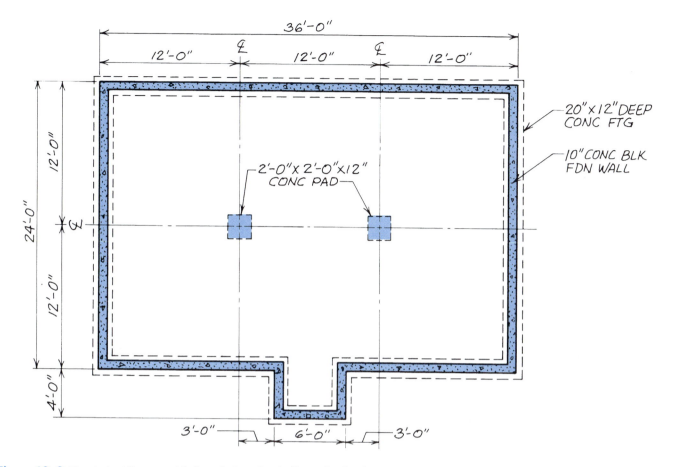

Figure 19–3 The dashed lines on this foundation plan indicate the footing.

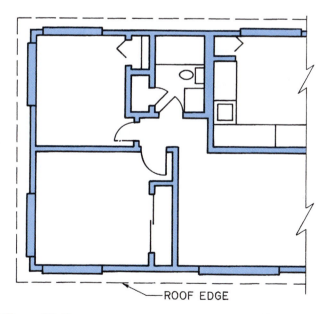

Figure 19–4 The dashed lines on this floor plan indicate the edge of the roof overhang above.

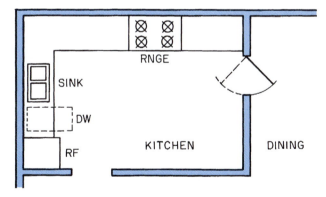

Figure 19–5 The dashed lines here are phantom lines to show alternate positions of the double-acting door and the door of the dishwasher.

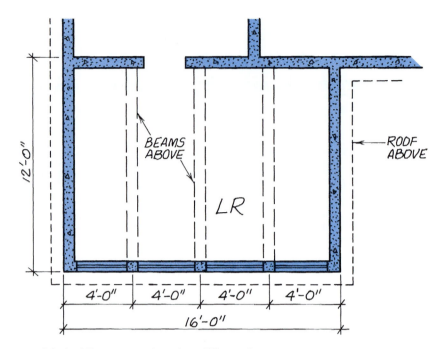

Figure 19–6 Different types of dashed lines are used to show different features.

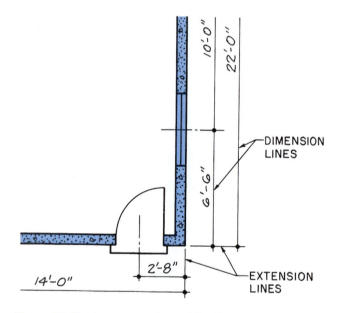

Figure 19–7 Dimension and extension lines.

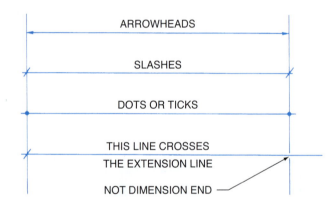

Figure 19–8 Dimension-line ends.

Extension Lines and Dimension Lines

Extension lines are thin, solid lines that project from an object to show the extent or limits of a dimension. Extension lines should not quite touch the object they indicate (Fig. 19–7).

Dimension lines are solid lines of the same weight as extension lines. A dimension line is drawn from one extension line to the next. The dimension (distance between the extension lines) is lettered above the dimension line. On construction drawings, dimensions are expressed in feet and inches. The ends of dimension lines are drawn in one of three ways, as shown in Figure 19–8.

Dimensions that can be added together to come up with one overall dimension are called *chain dimensions*. The dimension lines for chain dimensions are kept in line as much as possible. This makes it easier to find the dimensions that must be added to find the overall dimension.

Centerlines

Centerlines are made up of long and short dashes. They are used to show the centers of round or cylindrical objects. Centerlines are also used to indicate that an object is *symmetrical,* or the same on both sides of the center (Fig. 19–9). To show the center of a round object, two centerlines are used so that the short dashes cross in the center (Fig. 19–10).

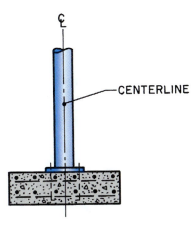

Figure 19–9 This centerline indicates that the column is symmetrical, or the same on both sides of the centerline.

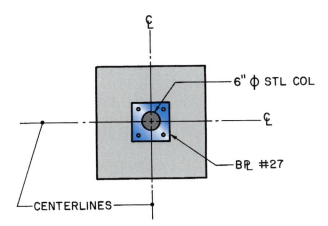

Figure 19–10 When centerlines show the center of a round object, the short dashes of the two centerlines cross.

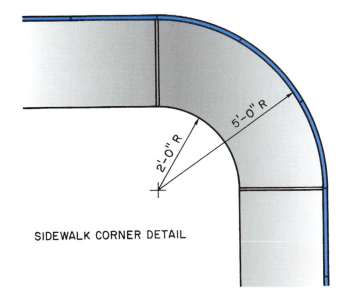

SIDEWALK CORNER DETAIL

Figure 19–11 Method of showing the radius of an arc.

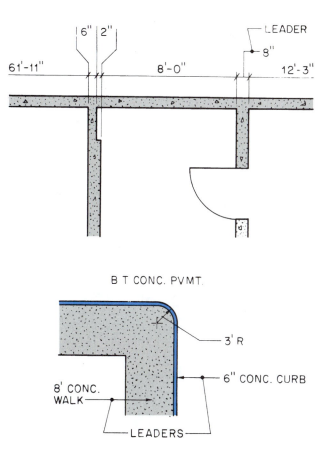

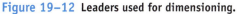

Figure 19–12 Leaders used for dimensioning.

To lay out an *arc* or part of a circle, the radius must be known. The *radius* of an arc is the distance from the center to the edge of the arc. On construction drawings, the center of an arc is shown by crossing centerlines. The radius is dimensioned on a thin line from the center to the edge of the arc (Fig. 19–11).

Rather than clutter the drawing with unnecessary lines, only the short, crossing dashes of the centerlines are shown. If the centerlines are needed to dimension the location of the center, only the needed centerlines are extended.

Leaders

Some construction details are too small to allow enough room for clear dimensioning by the methods described earlier. To overcome this problem, the dimension is shown in a clear area of the drawing. A thin line called a **leader** shows where the dimension belongs (Fig. 19–12).

Cutting-Plane Lines

It was established earlier that section views are needed to show interior detail. In order to show where the imaginary cut was made, a **cutting-plane line** is drawn on the view through which the cut was made (Fig. 19–13). A cutting-plane line is usually a heavy line with long dashes and pairs of short dashes. Some drafters, however, use a solid, heavy line. In either case, cutting-plane lines always have some identification at their ends. Cutting-plane-line identification symbols are discussed in the next chapter.

Some section views may not be referenced by a cutting-plane line on any other view. These are *typical sections* that would be the same if drawn from an imaginary cut in any part of the building (Fig. 19–14).

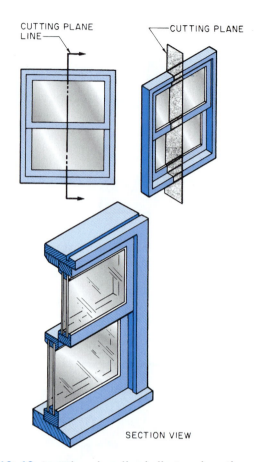

Figure 19–13 A cutting-plane line indicates where the imaginary cut is made and how it is viewed.

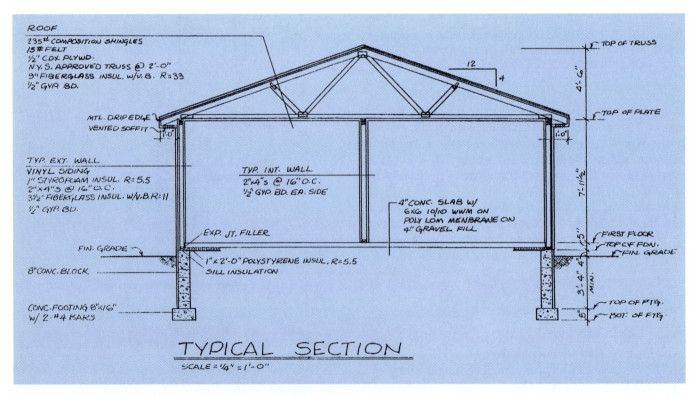

Figure 19–14 Building section. *Courtesy of Kurzon Architects.*

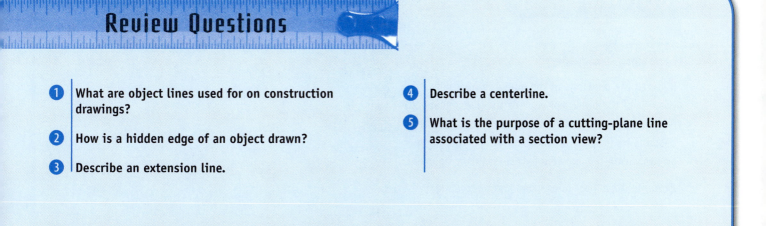

Review Questions

1. What are object lines used for on construction drawings?

2. How is a hidden edge of an object drawn?

3. Describe an extension line.

4. Describe a centerline.

5. What is the purpose of a cutting-plane line associated with a section view?

Activities

IDENTIFYING LINES

Refer to the drawings of the house in Appendix B at the end of the book. For each of the lines numbered A1 through A10, identify the kind of line and briefly describe its purpose on these drawings. The broad arrows with A numbers are for use in this assignment.

Example A0, *object line, shows end of building.*

Chapter 20 | Use of Symbols

OBJECTIVES

After completing this chapter, the student will be able to identify and understand the meaning of:

- ❌ door and window symbols.
- ❌ materials symbols.
- ❌ electrical and mechanical symbols.
- ❌ reference marks for coordinating drawings.
- ❌ abbreviations.

Glossary of Terms

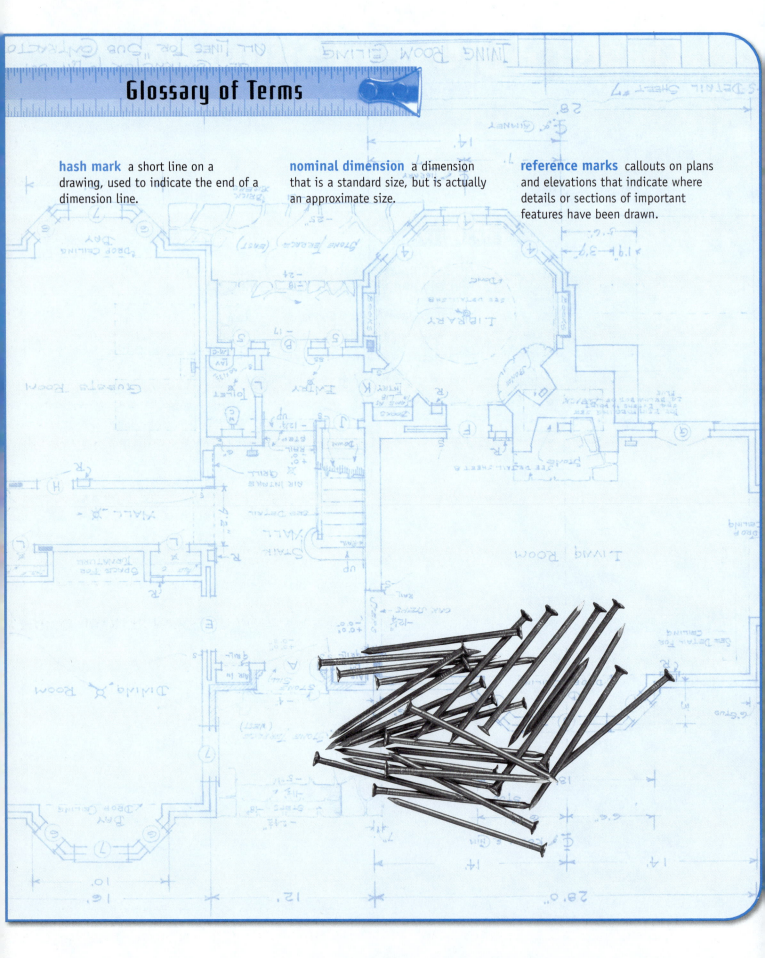

hash mark a short line on a drawing, used to indicate the end of a dimension line.

nominal dimension a dimension that is a standard size, but is actually an approximate size.

reference marks callouts on plans and elevations that indicate where details or sections of important features have been drawn.

An alphabet of lines allows for clear communication through drawings, and the use of standard symbols makes for even better communication. Many features of construction cannot be drawn exactly as they appear on the building. Therefore, standard symbols are used to show various materials, plumbing fixtures and fittings, electric devices, windows, doors, and other common objects. Notes are added to drawings to give additional explanations.

It is not important to memorize all the symbols and abbreviations used in construction before you learn to read drawings. You should, however, memorize a few of the most common symbols and abbreviations so that you may learn the principles involved in their use. Additional symbols and abbreviations can be looked up as they are needed. The illustrations shown here represent only a few of the more common symbols and abbreviations.

Door and Window Symbols

Door and window symbols show the type of door or window used and the direction the door or window opens. There are three basic ways for household doors to open—swing, slide, or fold (Fig. 20–1). Within each of these basic types there are variations that can be readily understood from their symbols. The direction a swing-type door opens is shown by an arc representing the path of the door.

There are seven basic types of windows. They are named according to how they open (Fig. 20–2). The symbols for hinged windows—awning, casement, and hopper—indicate the direction they open. In elevation, the symbols include dashed lines that come to a point at the hinged side, as viewed from the exterior.

The sizes of windows and doors are usually shown on a special window schedule or door schedule, but they might also be indicated by notes on the plans near their symbols. A door schedule or a window schedule is a table showing the type of door or window, rough opening sizes, and any other important information for the installation of doors and windows. The notations of size show width first and height second. Manufacturers' catalogs usually list several sets of dimensions for every window model (Fig. 20–3). The glass size indicates the area that will actually allow light to pass. The rough opening size is important for the carpenter who will frame the wall into which the window will be installed. The masonry opening is important to masons. The notations on plans and schedules usually indicate nominal dimensions. A **nominal dimension** is an approximate size and may not represent any of the actual dimensions of the unit. Nominal dimensions are usually rounded off to whole inches or to feet and inches and are used only as a convenient way to refer to the window or door size. The actual dimensions

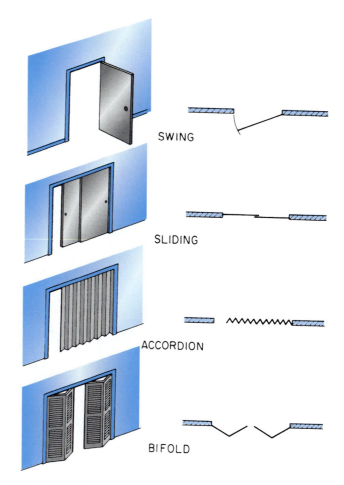

Figure 20–1 **Types of doors and their plan symbols.**

should be obtained from the manufacturer before construction begins.

Materials Symbols

The drawing of an object shows its shape and location. The outline of the drawing may be filled in with a material symbol to show what the object is made of (Fig. 20–4). Many materials are represented by one symbol in elevations and another symbol in sections. Examples of such symbols are concrete block and brick. Other materials look pretty much the same when viewed from any direction, and so their symbols are drawn the same in sections and elevations.

When a large area is made up of one material, it is common to only draw the symbol in a part of the area (Fig. 20–5). Some drafters simplify this even further by using a note to indicate what material is used and omitting the symbol altogether.

PLAN   ELEVATION   PICTORIAL

DOUBLE HUNG

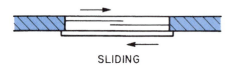

SLIDING

AWNING

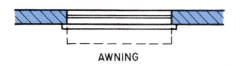

CASEMENT

HOPPER

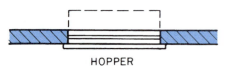

JALOUSIE

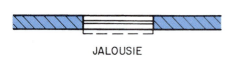

FIXED

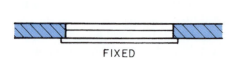

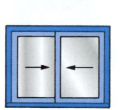

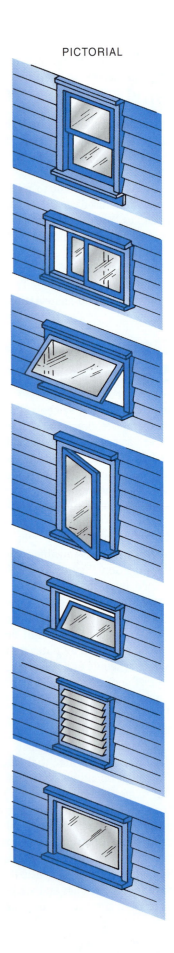

Figure 20–2 Window symbols.

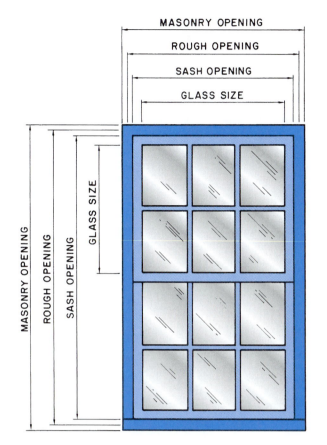

Figure 20–3 Windows and doors can be measured in several ways.

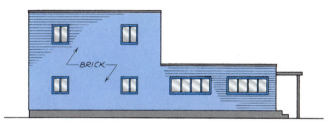

Figure 20–5 Only part of the area is covered by the brick symbol although the entire building will be brick.

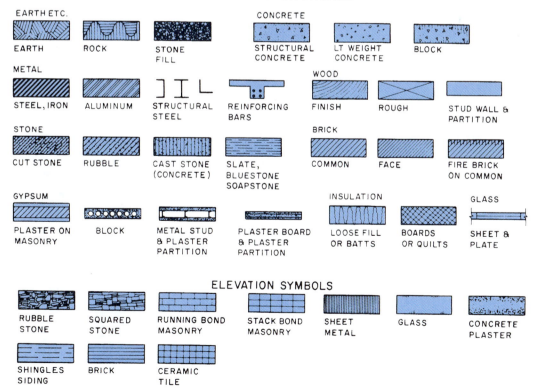

Figure 20–4 Material symbols.

Electrical and Mechanical Symbols

The electrical and mechanical systems in a building include wiring, electrical devices, piping, pipe fittings, plumbing fixtures, registers, and heating and air-conditioning ducts. It is not practical to draw these items as they would actually appear, and so standard symbols have been devised to indicate them.

The electrical system in a house includes wiring as well as devices such as switches, receptacles, light fixtures, and appliances. Wiring is indicated by lines that show how devices are connected. These lines are not shown in their actual position. They simply indicate which switches control which lights, for example. Outlets (receptacles) and switches are usually shown in their approximate positions. Major fixtures and appliances are shown in their actual positions. A few of the most common electrical symbols are shown in Figure 20-6.

Mechanical systems—plumbing and HVAC (heating, ventilating, and air conditioning)—are not usually shown in much detail on drawings for single-family homes. However, some of the most important features may be shown. Piping is shown by lines; different types of lines represent different kinds of piping. Symbols for pipe fittings are the same basic shape as the fittings they represent. A short line, or **hash mark,** represents the joint between the pipe and the fitting. Plumbing fixtures are drawn pretty much as the actual fixture appears. A few plumbing symbols are shown in Figure 20-7.

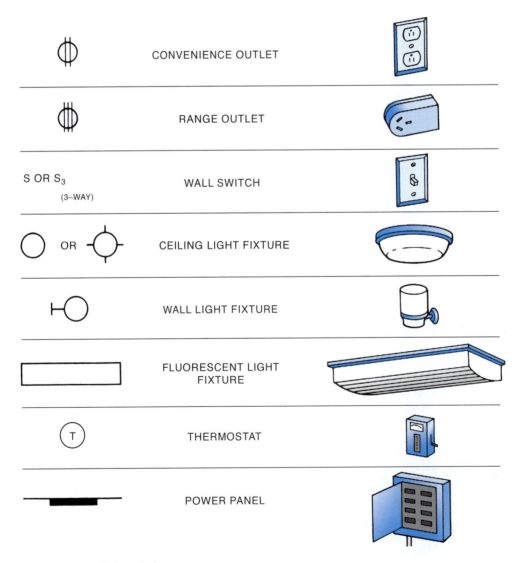

⊕	CONVENIENCE OUTLET
⊕	RANGE OUTLET
S OR S₃ (3–WAY)	WALL SWITCH
○ OR ⊕	CEILING LIGHT FIXTURE
⊢○	WALL LIGHT FIXTURE
▭	FLUORESCENT LIGHT FIXTURE
ⓣ	THERMOSTAT
▬	POWER PANEL

Figure 20–6 Some common electrical symbols.

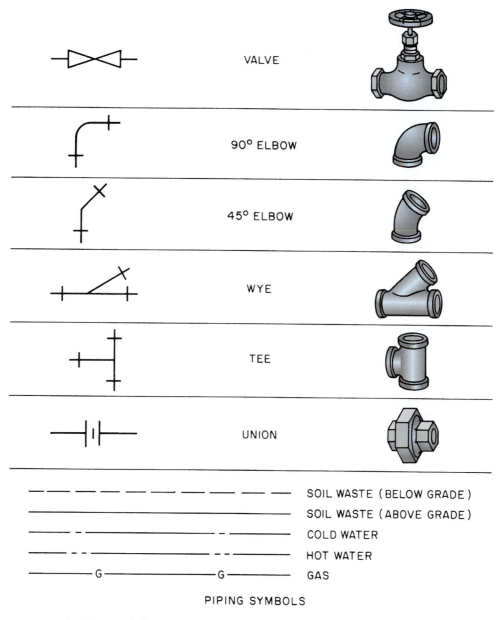

PIPING SYMBOLS

Figure 20–7 Some common plumbing symbols.

Reference Marks

A set of drawings for a complex building may include several sheets of section and detail drawings. These sections and details do not have much meaning without some way of knowing what part of the building they are meant to show. Callouts, called **reference marks,** on plans and elevations indicate where details or sections of important features have been drawn. To be able to use these reference marks for coordinating drawings, you must first understand the numbering system used on the drawings. The simplest numbering system for drawings consists of numbering the drawing

sheets and naming each of the views. For example, Sheet 1 might include a site plan and foundation plan; Sheet 2, floor plans; and Sheet 3, elevations.

On large, complex sets of drawings, the sheets are numbered according to the kind of drawings shown. Architectural drawing sheets are numbered A-1, A-2, and so on for all the sheets. Electrical drawings are numbered E-1, E-2, and E-3. A view number identifies each separate drawing or view on the sheet. Figure 20–8 shows drawing 5 on sheet A-4.

Because most of the drawings for a house are architectural, and the drawing set is fairly small, letters indicating the type of drawing are not usually included. Instead, the

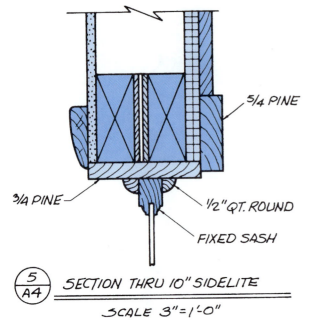

5/4 PINE

3/4 PINE

1/2" QT. ROUND

FIXED SASH

5/A4 SECTION THRU 10" SIDELITE

SCALE 3"=1'-0"

Figure 20-8 This is drawing 5 on sheet A-4.

views are numbered, and a second number shows on which sheet it appears. For example, the fourth drawing on the third sheet would be 4/3, 4.3, or 4-3.

Numbering each view and the sheet on which it appears makes it easy to reference a section or detail to another drawing. The identification of a section view is given with the cutting-plane line showing where it is taken from. For example, the section view in Figure 20-9 shows the fireplace at the cutting-plane line in Figure 20-10. Notice that the cutting-plane line in Figure 20-10 indicates that the section is viewed from the top of the page toward the bottom, with the fireplace opening on the right. That is how the section view in Figure 20-9 is drawn. This numbering system is also used for details that cannot be located by a cutting-plane line. The detail drawing of the cornice (edge of the roof) in Figure 20-11 is drawing 4 on sheet A-4 in Figure 20-12.

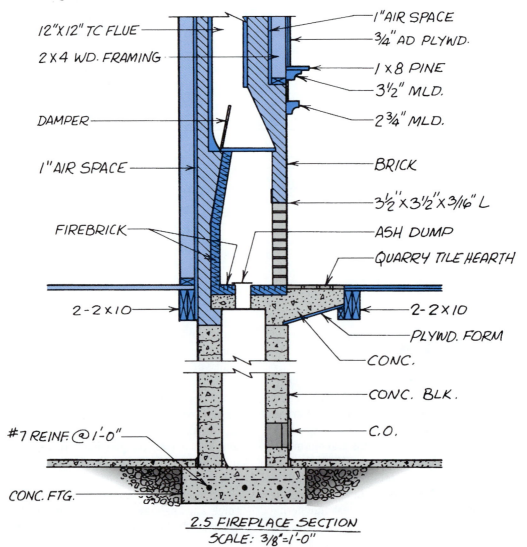

12"X12" TC FLUE

2X4 WD. FRAMING

DAMPER

1" AIR SPACE

FIREBRICK

2-2X10

#7 REINF. @ 1'-0"

CONC. FTG.

1" AIR SPACE

3/4" AD PLYWD.

1 X 8 PINE

3 1/2" MLD.

2 3/4" MLD.

BRICK

3 1/2" X 3 1/2" X 3/16" L

ASH DUMP

QUARRY TILE HEARTH

2-2X10

PLYWD. FORM

CONC.

CONC. BLK.

C.O.

2.5 FIREPLACE SECTION
SCALE: 3/8"=1'-0"

Figure 20-9 This section view is drawing 2 on sheet 5.

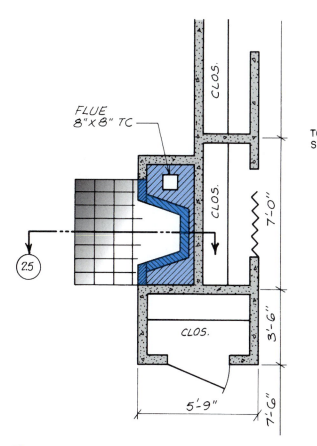

Figure 20–10 Plan for fireplace detailed in Figure 20–9.

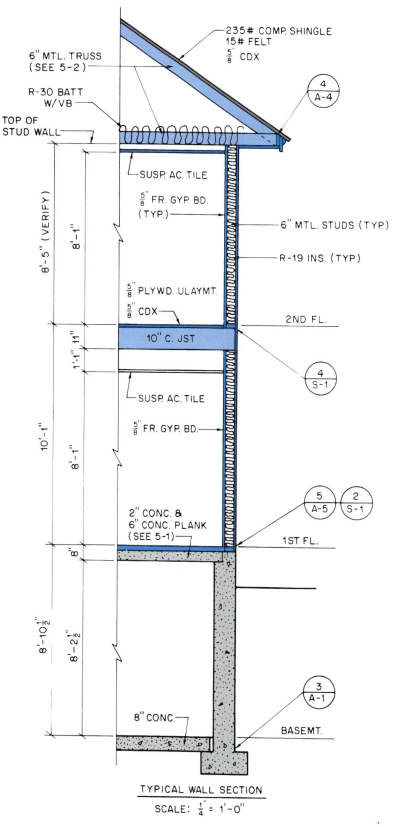

TYPICAL WALL SECTION

SCALE: $\frac{1}{4}$" = 1'-0"

Figure 20–11 The detail of this cornice is shown in drawing 4 on sheet A-4 of Figure 20–12.

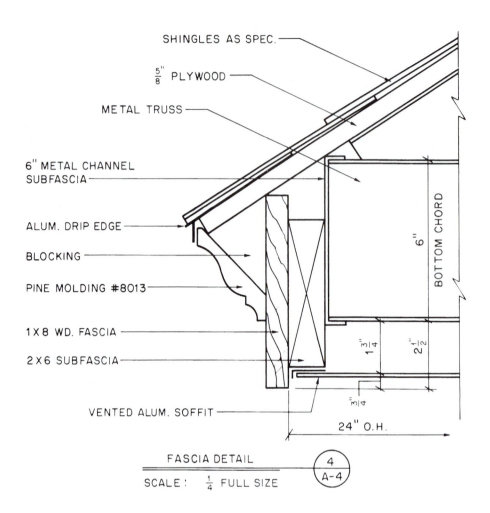

SHINGLES AS SPEC.

$\frac{5}{8}$" PLYWOOD

METAL TRUSS

6" METAL CHANNEL
SUBFASCIA

ALUM. DRIP EDGE

BLOCKING

PINE MOLDING #8013

1 X 8 WD. FASCIA

2 X 6 SUBFASCIA

BOTTOM CHORD

6"

$1\frac{3}{4}$" $2\frac{1}{2}$"

$\frac{3}{4}$"

VENTED ALUM. SOFFIT

24" O.H.

FASCIA DETAIL

SCALE : $\frac{1}{4}$ FULL SIZE

4
A-4

Figure 20–12 This is the detail of the cornice in Figure 20–11.

Abbreviations

Drawings for construction include many notes and labels of parts. These notes and labels are abbreviated as much as possible to avoid crowding the drawing. The abbreviations used on drawings are usually a shortened form of the word and are easily understood. For example, BLDG stands for building. Figure 20–13 lists common abbreviations.

Abbreviations

A.B.—anchor bolt
A.C.—air conditioning
AL. or ALUM.—aluminum
BA—bathroom
BLDG.—building
BLK.—block
BLKG.—blocking
BM.—beam
BOTT.—bottom
B.PL.—base plate
BR—bedroom
BRM.—broom closet
BSMT.—basement
CAB.—cabinet
℄—centerline
CLNG. or CLG.—ceiling
C.M.U.—concrete masonry unit
    (concrete block)
CNTR.—center or counter
COL.—column
COMP.—composition
CONC.—concrete
CONST.—construction
CONT.—continuous
CORRUG.—corrugated
CRNRS.—corners
CU—copper
d—penny (nail size)
DBL.—double
DET.—detail
DIA or ∅—diameter
DIM.—dimension
DN.—down
DO—ditto
DP.—deep or depth
DR.—door
D.W.—dishwasher
ELEC.—electric
ELEV.—elevation
EQ.—equal

EXP.—exposed or expansion
EXT.—exterior
F.G.—fuel gas
FIN.—finish
FL. or FLR.—floor
FOUND. or FDN.—foundation
F.P.—fireplace
FT.—foot or feet
FTG.—footing
GAR.—garage
G.F.I.—ground fault interrupter
G.I.—galvanized iron
GL.—glass
GRD.—grade
GYP.BD.—gypsum board
H.C.—hollow core door
H.C.W.—hollow core wood
HDR.—header
H.M.—hollow metal
HORIZ.—horizontal
HT. or HGT.—height
H.W.—hot water
H.W.M.—high water mark
IN.—inch or inches
INSUL.—insulation
INT.—interior
JSTS.—joists
JT.—joint
LAV.—lavatory
L.H.—left hand
LIN.—linen closet
LT.—light
MANUF.—manufacturer
MAS.—masonry
MATL.—material
MAX.—maximum
MIN.—minimum
MTL.—metal
NAT.—natural
N/F—now or formerly

N.I.C.—not in contract
o/—overhead or over
O.C.—on centers
O.H. DOOR—overhead door
PERF.—perforated
℞—plate
PLYWD.—plywood
P.T.—pressure-treated lumber
R—risers
REF.—refrigerator
REINF.—reinforcement
REQ.—requirement
R.H.—right hand
RM—room
R.O.B.—run of bank (gravel)
R.O.W.—right of way
SCRND.—screened
SHT.—sheet
SHTG.—sheathing
SHWR.—shower
SIM.—similar
SL.—sliding
S&P—shelf and pole
SQ. or ⊡—square
STD.—standard
STL.—steel
STY.—story
T&G—tongue and groove
THK.—thick
T'HOLD.—threshold
TYP.—typical
V.B.—vapor barrier
w/—with
WARD.—wardrobe
W.C.—water closet
WD.—wood
WDW.—window
W.H.—water heater
W.I.—wrought iron

Figure 20–13 **Common abbreviations found on drawings.**

Review Questions

1 **What is represented by each of these symbols?**

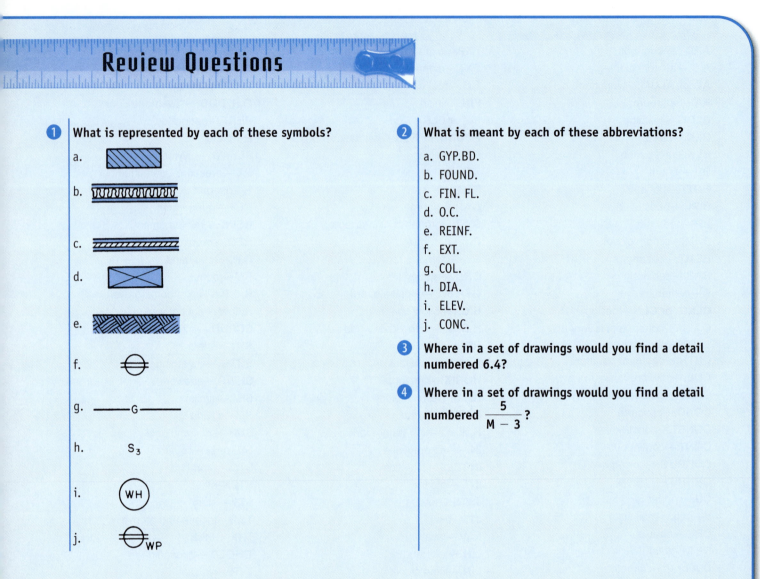

a.

b.

c.

d.

e.

f.

g. ——G——

h. S_3

i. WH

j. WP

2 **What is meant by each of these abbreviations?**

a. GYP.BD.

b. FOUND.

c. FIN. FL.

d. O.C.

e. REINF.

f. EXT.

g. COL.

h. DIA.

i. ELEV.

j. CONC.

3 **Where in a set of drawings would you find a detail numbered 6.4?**

4 **Where in a set of drawings would you find a detail numbered $\dfrac{5}{M-3}$?**

Activities

READING SYMBOLS

Before the contracts are even awarded for the construction of a house, the contractor must estimate the cost of all the materials and their installation. Part of this estimating process is a review of the plans, counting symbols for various devices. In this activity, you are to review the drawings in Appendix B at the end of the book and answer the following questions:

1. How many duplex electric outlets are shown on the floor plan?
2. How many light fixtures are shown on the floor plan?
3. How many single-pole light switches are shown on the floor plan?
4. How many three-way light switches are shown on the floor plan?
5. How many windows are shown on the floor plan?
6. What material is on the outside of the wall near the tub at the front of the house?
7. According to the foundation plan, what is the basement floor below bedrooms 1 and 2 made of?
8. According to the foundation plan, what is the basement floor below the activity room made of?

Chapter 21 | Plan Views

OBJECTIVES

After completing this chapter, the student will be able to explain the general kinds of information shown on:

- ⊗ site plans.
- ⊗ foundation plans.
- ⊗ floor plans.

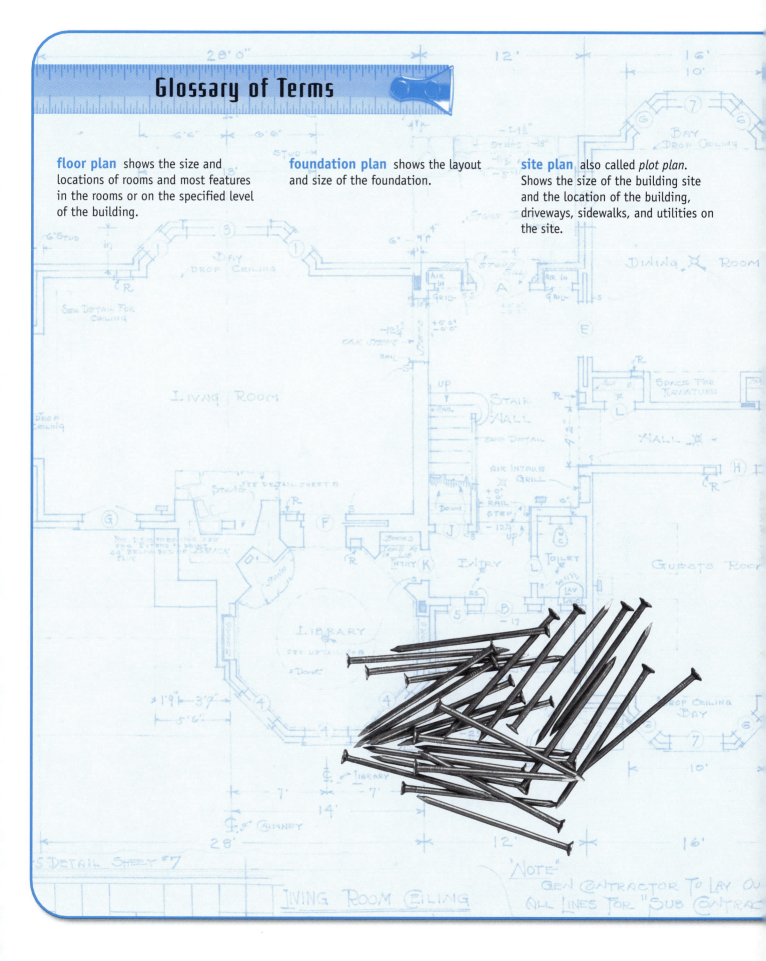

Glossary of Terms

floor plan shows the size and locations of rooms and most features in the rooms or on the specified level of the building.

foundation plan shows the layout and size of the foundation.

site plan also called *plot plan*. Shows the size of the building site and the location of the building, driveways, sidewalks, and utilities on the site.

ou learned in Chapter 17 that plans are drawings that show an object as viewed from above. Many of the detail and section drawings in a set show parts of the building from above. Some of the plan views that show an entire building are discussed here. This brief explanation will help you feel more comfortable with plans, although it does not cover plans in depth. You will use plans frequently throughout your study of construction.

Site Plans

A **site plan** gives information about the site on which the building is to be constructed. The boundaries of the site (property lines) are shown. The property line is usually a heavy line with one or two short dashes between longer line segments. The lengths of the boundaries are noted next to the line symbol. Property descriptions are often the result of a survey by a surveyor or civil engineer. These professionals usually work with decimal parts of feet, rather than feet and inches. Therefore, site dimensions are usually stated in tenths or hundredths of feet (Fig. 21–1).

A north arrow of some type indicates what compass direction the site faces. Unless this north arrow includes a correction for the difference between true north and magnetic north, it may be only an approximation. However, it is sufficient to show the general direction the site faces.

The site plan also indicates where the building is positioned on the site. As a minimum, the dimensions to the front and one side of the site are given. The overall dimensions of the building are also included. Anyone reading the site plan will have this basic information without referring to the other drawings. If the finished site is to include walks, drives, or patios, these are also described by their overall dimensions.

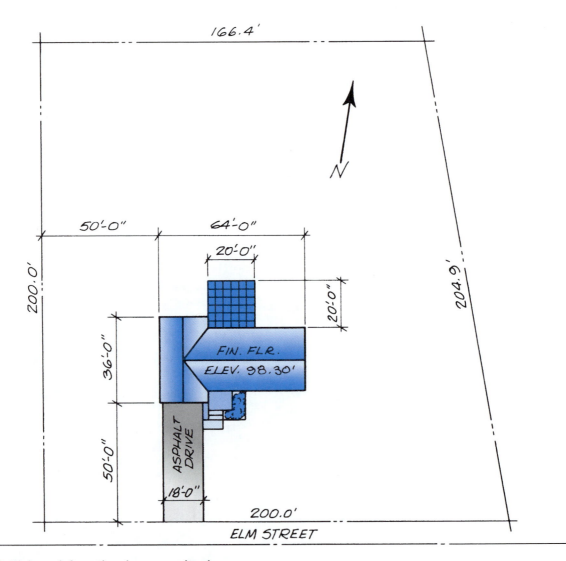

Figure 21–1 Minimum information shown on a site plan.

Foundation Plans

A **foundation plan** is like a floor plan, but shows the foundation instead of the living spaces. It shows the foundation walls and any other structural work to be done below the living spaces.

Two types of foundations are commonly used in homes and other small buildings. One type has a concrete base, called the *footing,* supporting foundation walls (Fig. 21–2). The other is the slab-on-grade type. A *slab-on-grade* foundation consists of a concrete slab placed directly on the soil with little or no other support. Slabs on grade are usually thickened at their edges and wherever they must support a heavy load (Fig. 21–3).

When the footing-and-wall type foundation is used, girders are used to provide intermediate support to the structure above (Fig. 21–4). The girder is shown on the foundation plan by phantom lines and a note describing it.

The foundation plan includes all the dimensions necessary to lay out the footings and foundation walls. The footings follow the walls and may be shown on the plan. If they are shown, it is usually by means of hidden lines to show their outline only. In addition to the layout of the foundation walls, dimensions are given for opening windows, doors, and ventilators. Notes on the plan indicate areas that are not to be excavated, concrete-slab floors, and other important information about the foundation (Fig. 21–5).

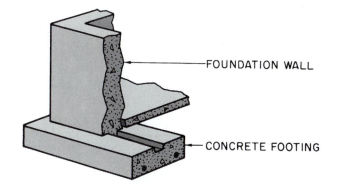

Figure 21–2 **Footing and foundation wall.**

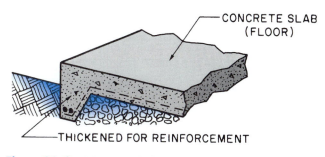

Figure 21–3 **Slab-on-grade foundation.**

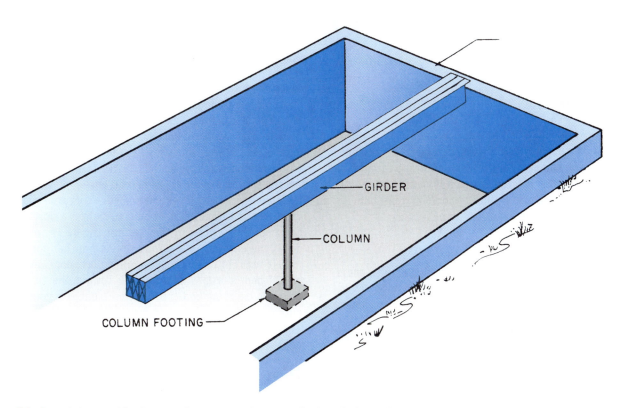

Figure 21–4 **A girder provides intermediate support between the foundation walls.**

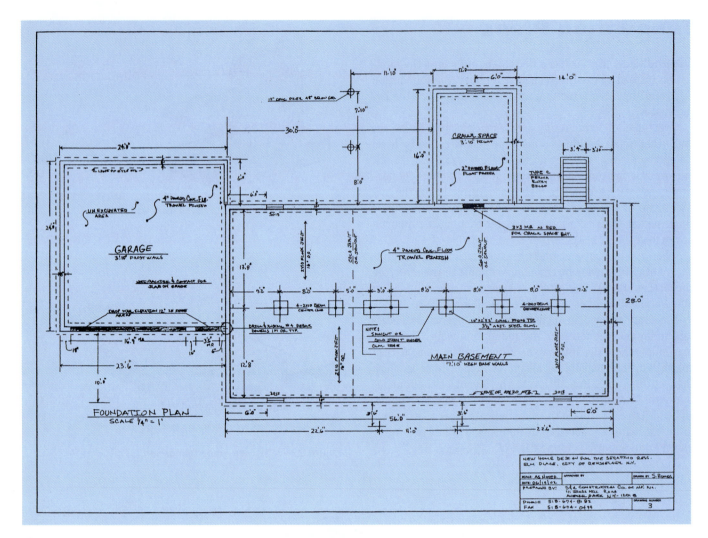

Figure 21–5 Foundation plan.

Floor Plans

A floor plan is similar to a foundation plan. It is a section view taken at a height that shows the placement of walls, windows, doors, cabinets, and other important features. A separate floor plan is included for each floor of the building. The floor plans provide more information about the building than any of the other drawings do.

Building Layout

The floor plans show the locations of all the walls, doors, and windows. Therefore, the floor plans show how the building is divided into rooms and how to get from one room to another. Before attempting to read any of the specific information on the floor plans, it is wise to familiarize yourself with the general layout of the building.

To quickly familiarize yourself with a floor plan, imagine that you are walking through the house. For example, imagine yourself standing in the front door of the house shown on the floor plan at the end of this textbook. As you enter the

foyer, the kitchen is on your right. Walk into the kitchen and the sink is on your right. On the far side of the kitchen is a laundry closet and to the right of the laundry closet is the door into the garage. Now turn around and proceed to the back of the house, where you enter the large activity room with a fireplace on the far wall. Turn left and you will see the door into the stairway at the far end of the room. Exit the activity room into the foyer and immediately turn right to walk down the hall. The first door on your right is a closet, the second goes into a powder room. Across the hall from the powder is the entrance to bedroom three. Another door on your right goes into bedroom one and the last door on your left enters bedroom two. If you walk toward the front of the house in bedroom two you will find a door on your left that enters a bathroom. Another door in the bathroom takes you into bedroom three. As you take this imaginary tour of the house, make note of as many details as possible.

Dimensions

Dimensions are given for the sizes and locations of all walls, partitions, doors, windows, and other important

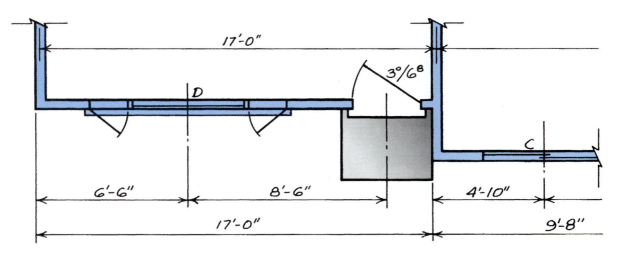

Figure 21–6 Frame construction dimensioning.

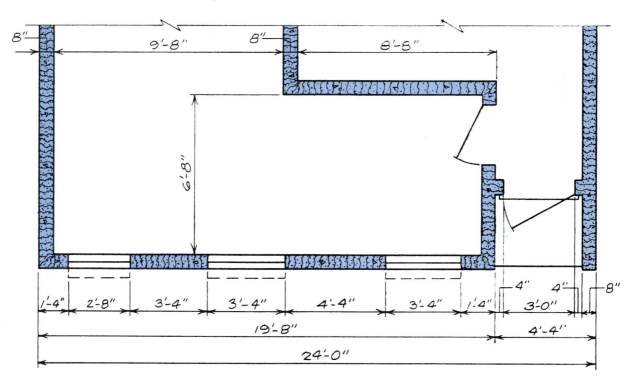

Figure 21–7 Masonry construction dimensioning.

features. On frame construction, exterior walls are usually dimensioned to the outside face of the wall framing. If the walls are to be covered with stucco or masonry veneer, this material is outside the dimensioned face of the wall frame. Interior partitions may be dimensioned to their centerlines or to the face of the studs. (*Studs* are the vertical members in a wall frame.) Windows and doors may be dimensioned about their centerlines (Fig. 21-6) or to the edges of the openings.

Solid masonry construction is dimensioned entirely to the face of the masonry (Fig. 21-7). Masonry openings for doors and windows are dimensioned to the edge of the openings.

Other Features of Floor Plans

The floor plan includes as much information as possible without making it cluttered and hard to read. Doors and windows are shown by their symbols as explained in Chapter 20. Cabinets are shown in their proper positions. The cabinets are explained further by cabinet elevations and details, which are discussed in Chapter 23. If the building includes stairs, these are shown on the floor plan. Important overhead construction is also indicated on the floor plans. If the ceiling is framed with joists, their size, direction, and spacing are shown on the floor plan. Architectural features such as exposed beams, arches in doorways, or unusual roof lines may be shown by phantom lines.

Review Questions

1 From what direction is an object viewed in a plan view?

2 What drawing in a set usually shows where the property boundaries are?

3 What is indicated by a bold arrow with the letter "N" next to it?

4 What drawing in a set shows the girder under the main floor if the building has a girder?

5 What drawing shows the locations of walls and the sizes of room?

6 For frame construction, are exterior walls usually dimensioned from the outside of the wall, the centerline of the wall, or the inside of the wall?

7 What drawing shows the locations of plumbing fixtures?

8 Are dimensions for concrete walls usually taken from the centerline of the wall or the face of the wall?

Activities

READING PLAN VIEWS

Refer to the drawings of the house in Appendix B at the end of the book to complete this assignment.

1. In what direction does the front of the house face?
2. What is the width of the lot at the front?
3. How far is the front of the house from the edge of the street?
4. What is the overall length and width of the house?
5. What are the inside dimensions of bedroom #2?
6. What is the thickness of the partitions between bedroom #3 and the hall?
7. What is the thickness of the interior wall between the powder room and the stairs?
8. How many windows are in the garage?
9. What is the distance from the west end of the house to the centerline of the front entrance?
10. What is the 4-inch-thick material on the outside of the wall around the bathtub at the front of the house?
11. What are the outside dimensions of the terrace?
12. What is the inside width of the linen closet at the end of the hall?

Chapter 22 Elevations

OBJECTIVES

After completing this chapter, the student will be able to:

- orient building elevations to building plans.
- explain the kinds of information shown on elevations.

256

Glossary of Terms

elevations drawings that show the height of objects.

grade line a line on a drawing that shows the surface level of the ground.

headers the framing across the top of a window or door opening.

orienting showing the relationship of one drawing to another.

plate the uppermost framing member in the wall.

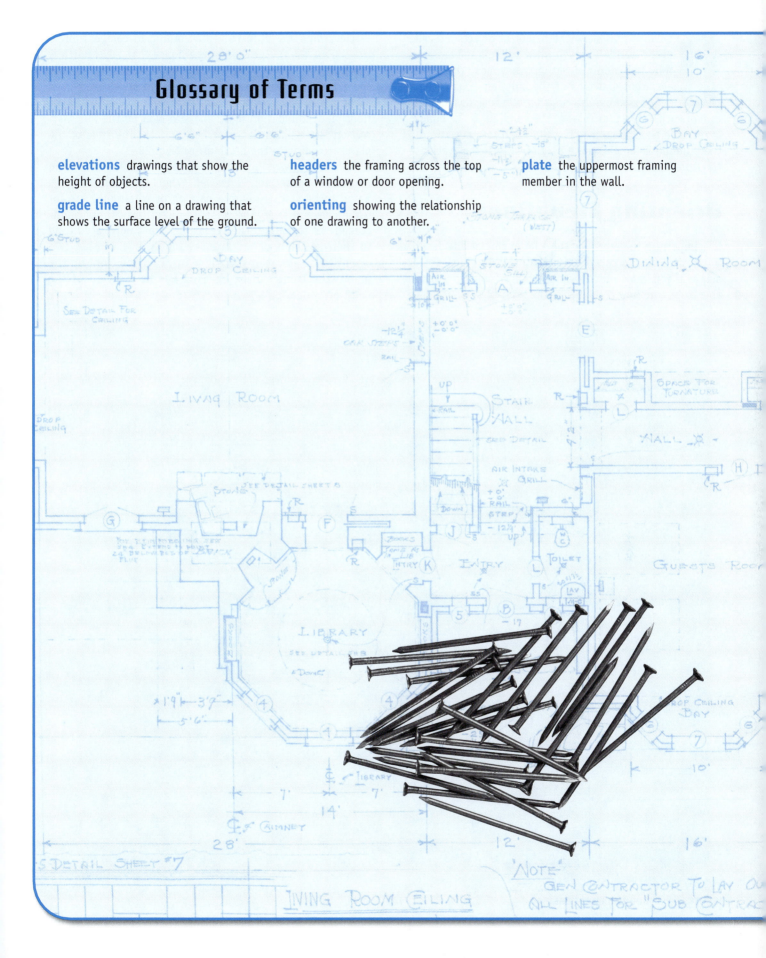

rawings that show the height of objects are called **elevations.** However, when builders and architects refer to building elevations, they mean the exterior elevation drawings of the building (Fig. 22–1). A set of working drawings usually includes an elevation of each of the four sides of the building. If the building is very complex, there may be more than four elevations. If the building is simple, there may be only two elevations—the front and one side.

Orienting Elevations

It is important to determine the relationship of one drawing to another. This is called **orienting** the drawings. For example, if you know which elevation is the front, you must be able to picture how it relates to the front of the floor plan.

Elevations are often named according to compass directions (Fig. 22–2). The side of the house that faces north is the north elevation, and the side that faces south is the south elevation, for example. When the elevations are named according to compass direction, they can be oriented to the floor plan, foundation plan, and site plan by the north arrow on those plans. It might help to label the edge of the plans according to the north arrow (Fig. 22–3).

It is not always possible to label elevations according to compass direction, however. When drawings are prepared to be sold through a catalog or when they are for use on several sites, the compass directions cannot be included. In this case, the elevations are named according to their position as you face the building (Fig. 22–4). To orient these elevations to the plans, find the front on the plans. The front can be checked by the location of the main entrance.

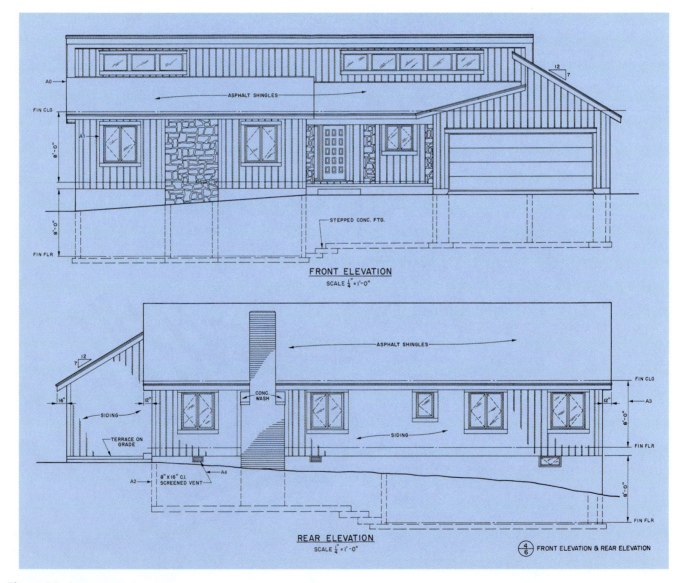

Figure 22–1 Building elevations.

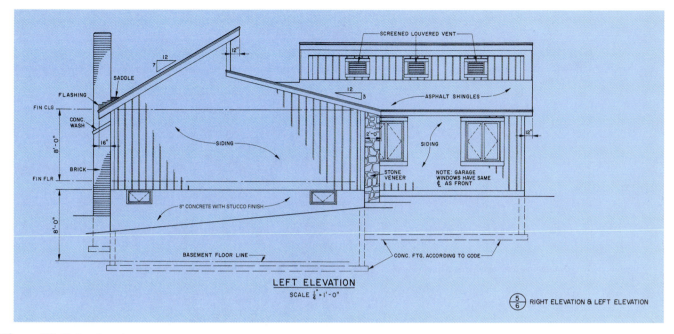

Figure 22–1 Continued.

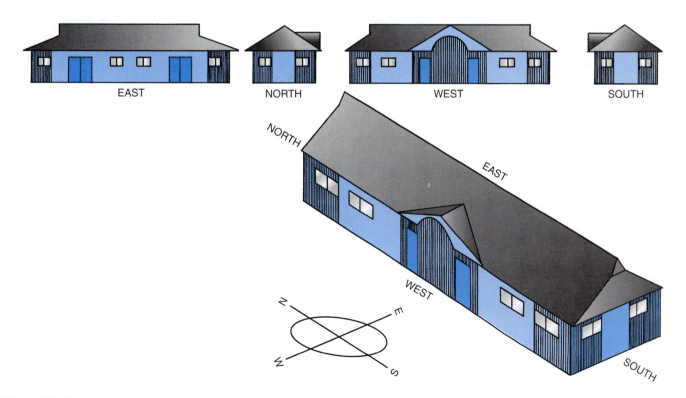

Figure 22–2 Elevations may be named according to their compass directions.

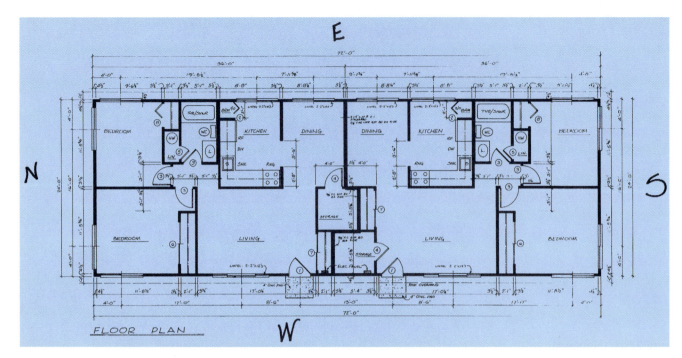

Figure 22–3 Plan labeled to help orientation to north arrow. *Courtesy of Kurzon Architects.*

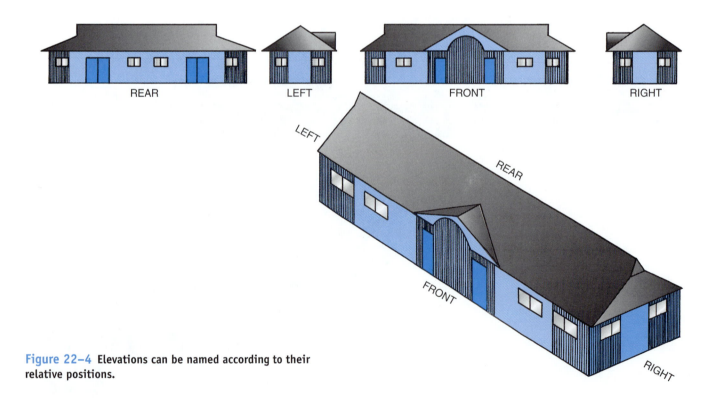

Figure 22–4 Elevations can be named according to their relative positions.

Information on Building Elevations

Building elevations are normally quite simple. Although the elevations do not include a lot of detailed dimensions and notes, they show the finished appearance of the building better than other views. Therefore, elevations are a great aid in understanding the rest of the drawing set.

The elevations show most of the building, as it will actually appear, with solid lines. However, the underground portion of the foundation is shown as hidden lines (Fig. 22–5). The footing is shown as a rectangle of dashed lines at the bottom of the foundation walls.

The surface of the ground is shown by a heavy solid line, called a **grade line.** The grade line usually includes one or more notes to indicate the elevation above sea level or an-

other reference point (Fig. 22–6). The word "elevation" used in this sense is altitude, or height—not a type of drawing. All references to the height of the ground or the level of key parts of the building are in terms of elevation.

Some important dimensions are included on the building elevations. Most of them are given in a string at the end of one or more elevations (Fig. 22–7). The dimensions most often included are:

- thickness of footing.
- height of foundation walls.
- top of foundation to finished first floor.
- finished floor to ceiling or top of plate. (The **plate** is the uppermost framing member in the wall.)
- finished floor to bottom of window headers. (The **headers** are the framing across the top of a window opening.)
- roof overhang at eaves.

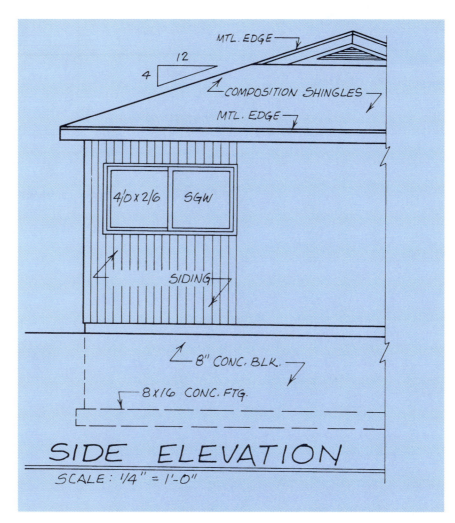

Figure 22–5 **The underground portion of the building is shown with dashed lines.** *Courtesy of Kurzon Architects.*

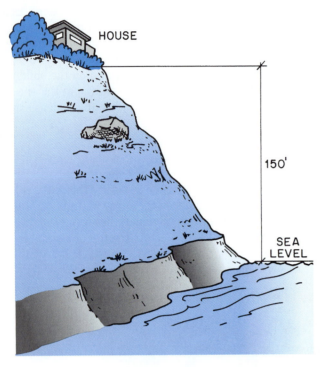

Figure 22–6 The elevation of this site is 150 feet.

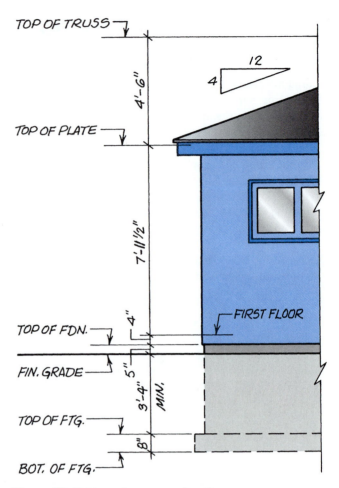

Figure 22–7 Dimensions on an elevation.

Review Questions

1 What is an elevation drawing?

2 What is indicated by a heavy solid line near the bottom of a building on an exterior elevation?

3 What is indicated by dashed lines near the bottom of a building on an exterior elevation?

4 List four dimensions that are usually shown on building elevations.

Activities

READING ELEVATIONS

Refer to the drawings of the house in Appendix B at the end of the book to complete this assignment.

1. Which elevation is the north elevation?
2. In what compass direction does the left end of the house face?
3. What is the dimension from the surface of the floor to the ceiling?
4. How far does the main house roof overhang beyond the walls?
5. What is the dimension from the surface of the basement floor to the bottom of the main floor framing?
6. What is the finish on the exposed portion of the foundation wall?
7. Above what space are the screened louvered vents in the left elevation?
8. How can you determine how deep in the ground to place the footings?

Sections and Details

After completing this chapter, the student will be able to:

- find and explain information shown on section views.
- find and explain information shown on large-scale details.
- orient sections and details to the other plans and elevations.

265

Glossary of Terms

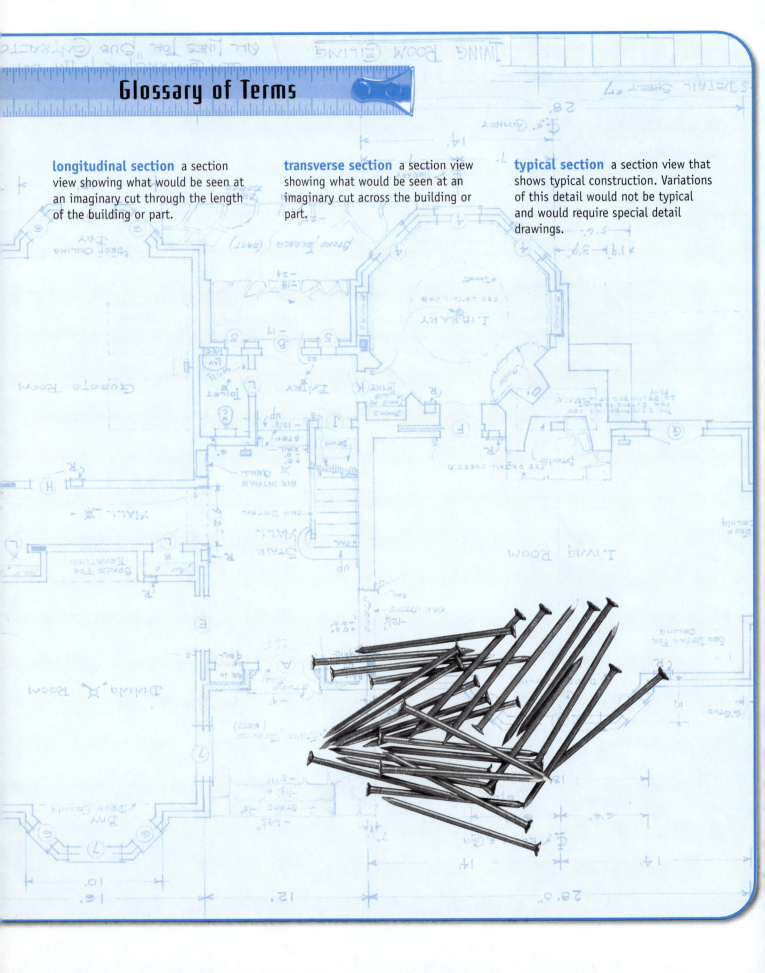

longitudinal section a section view showing what would be seen at an imaginary cut through the length of the building or part.

transverse section a section view showing what would be seen at an imaginary cut across the building or part.

typical section a section view that shows typical construction. Variations of this detail would not be typical and would require special detail drawings.

t is not possible to show all the details of construction on foundation plans, floor plans, and building elevations. Those drawings are meant to show the relationships of the major building elements to one another. To show how individual pieces fit together, it is necessary to use larger-scale drawings and section views. These drawings are usually grouped together in the drawing set. They are referred to as *sections and details* (Fig. 23–1).

Sections

Nearly all sets of drawings include, at least, a typical wall section. The **typical section** may be a section view of one wall, or it may be a full section of the building. Full sections are named by the direction in which the imaginary cut is made. Figure 23–2 shows a transverse section. A **transverse section** is taken from an imaginary cut across the width of the building. Transverse sections are sometimes called *cross sections*. A full section taken from a lengthwise cut through the building is called a **longitudinal section** (Fig. 23–3).

Full sections and wall sections normally have only a few dimensions, but have many notes with leaders to identify the parts of the wall. The following is a list of the kinds of information included on typical wall sections with most sets of drawings:

- Footing size and material (This may be specified by building codes.)
- Foundation wall thickness, height, and material
- Insulation, waterproofing, and interior finish for foundation walls
- Fill and waterproofing under concrete floors
- Concrete floor thickness, material, and reinforcement
- Sizes of floor framing materials
- Sizes of wall framing materials
- Wall covering (sheathing, siding, stucco, masonry, and interior wall finish) and insulation
- Cornice construction—materials and sizes (The *cornice* is the construction at the roof eaves.)
- Ceiling construction and insulation

Other section drawings are included as necessary to explain special features of construction. Wherever wall construction varies from the typical wall section, another wall section should be included. Section views are used to show any special construction that cannot be shown on normal plans and elevations. Figure 23–4 is an example of a special

section in elevation. This section view is said to be *in elevation* because it shows the height of the ridge construction. Figure 23–5 is *in plan* because it shows the interior of the fireplace as viewed from above.

Other Large-Scale Details

Sometimes necessary information can be conveyed without showing the interior construction. A large scale may be all that is needed to show the necessary details. The most common examples of this are on cabinet installation drawings (see Figure 23–6). Cabinet elevations show where the cabinets are located, without showing the interior construction.

Many details are best shown by combining elevations and sections or by using isometric drawings. Figure 23–7 shows an example of an elevation and a section used together to explain the construction of a fireplace. Figure 23–8 shows an isometric detail drawing that includes sections to show interior construction. Another method of showing detail is an exploded view. Figure 23–9 shows an exploded view of an electric receptacle.

Orienting Sections and Details

As explained earlier, some sections and details are labeled as typical. These drawings describe the construction that is used throughout most of the building.

Details and sections that refer to only one place in the building are identified by a reference mark. As was pointed out earlier in Chapter 19, sections are usually referenced by a cutting-plane line. This line shows where the section was taken from. Arrows on the ends of the cutting-plane line indicate what direction the imaginary cut is viewed from. A reference mark near the arrow indicates where the detail drawing is shown. These reference marks used for orienting details may vary from one set of drawings to another. It is important, although not usually difficult, to study the drawings and learn how the architect references details. Normally a system of sheet numbers and view numbers is used. One such numbering system was explained earlier.

Some basic principles of details and sections have been discussed here. You will gain more practice later in reading details and sections.

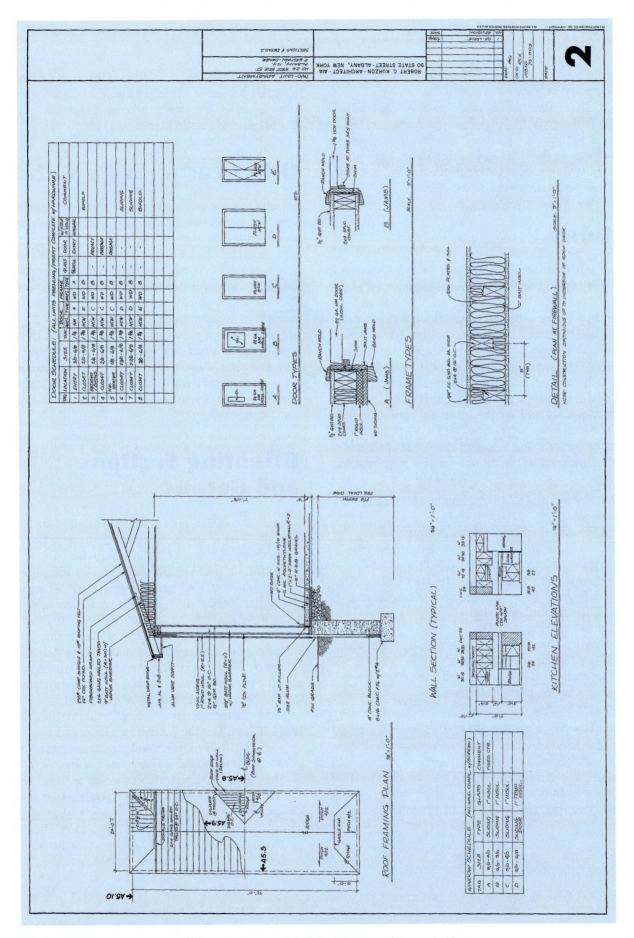

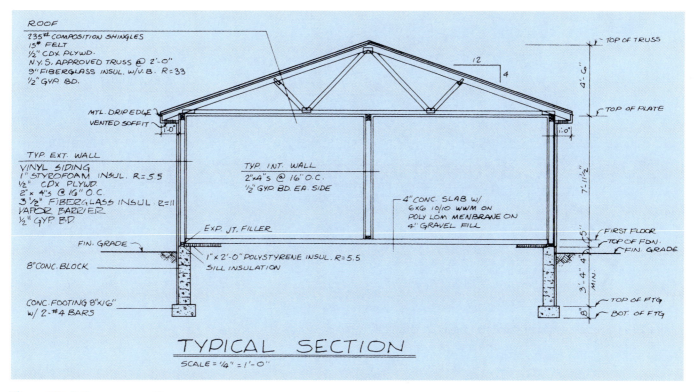

ROOF
235# COMPOSITION SHINGLES
15# FELT
½" CDX PLYWD.
N.Y.S. APPROVED TRUSS @ 2'-0"
9" FIBERGLASS INSUL. W/V.B. R=33
½" GYP BD.

MTL. DRIP EDGE
VENTED SOFFIT

TYP. EXT. WALL
VINYL SIDING
1" STYROFOAM INSUL. R=5.5
½" CDX PLYWD.
2 x 4's @ 16" O.C.
3 ½" FIBERGLASS INSUL. R=11
VAPOR BARRIER
½" GYP BD.

TYP. INT. WALL
2"x4"s @ 16" O.C.
½" GYP BD. EA. SIDE

4" CONC. SLAB W/
6x6 10/10 WWM ON
POLY LOM MENBRANE ON
4" GRAVEL FILL

TOP OF TRUSS
TOP OF PLATE

EXP. JT. FILLER

FIN. GRADE

8" CONC. BLOCK

1" x 2'-0" POLYSTYRENE INSUL. R=5.5
SILL INSULATION

FIRST FLOOR
TOP OF FDN.
FIN. GRADE

CONC. FOOTING 8"x16"
W/ 2-#4 BARS

TOP OF FTG
BOT. OF FTG

12
4

0'-0"
7'-11½"
5"
3'-4" MIN.
4"
8"

TYPICAL SECTION
SCALE = ¼" = 1'-0"

Figure 23–2 Transverse section.

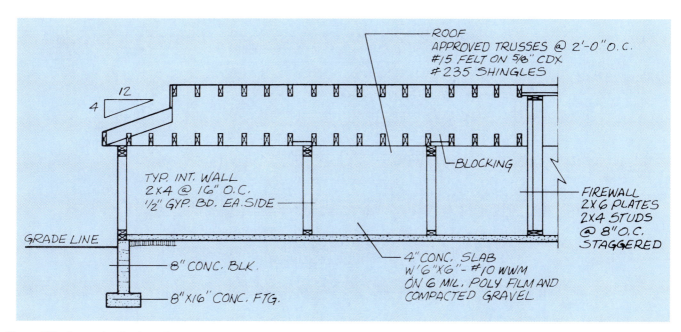

ROOF
APPROVED TRUSSES @ 2'-0" O.C.
#15 FELT ON ⅝" CDX
#235 SHINGLES

12
4

TYP. INT. WALL
2X4 @ 16" O.C.
½" GYP. BD. EA. SIDE

BLOCKING

FIREWALL
2X6 PLATES
2X4 STUDS
@ 8" O.C.
STAGGERED

GRADE LINE

8" CONC. BLK.

8"X16" CONC. FTG.

4" CONC. SLAB
W/ 6"X6" - #10 WWM
ON 6 MIL. POLY FILM AND
COMPACTED GRAVEL

Figure 23–3 Longitudinal section.

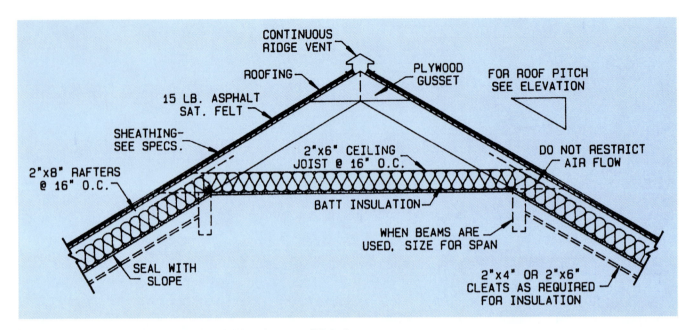

Figure 23–4 Special section of ventilated ridge. *Courtesy of W. D. Farmer.*

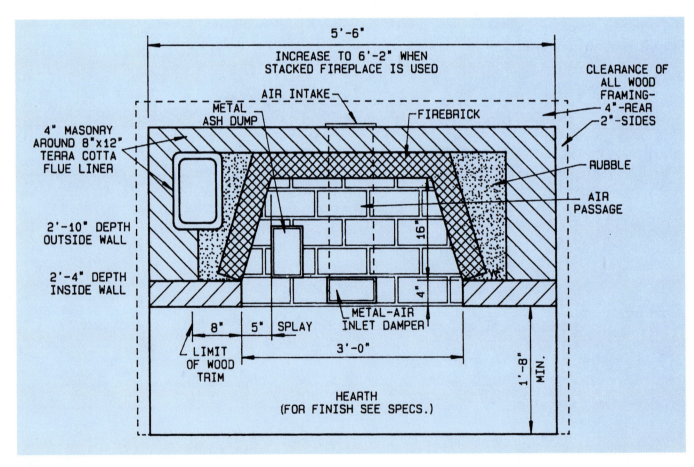

Figure 23–5 A section in plan. *Courtesy of W. D. Farmer.*

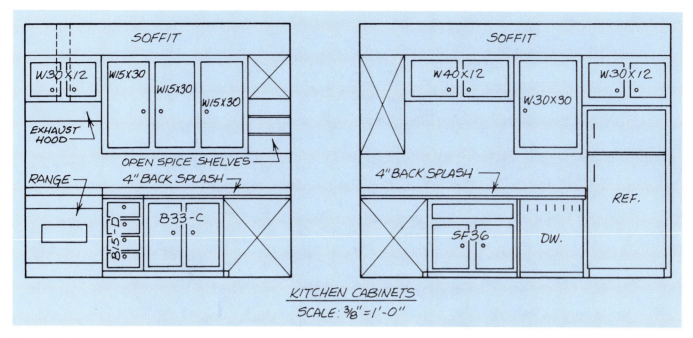

Figure 23–6 Cabinet elevations.

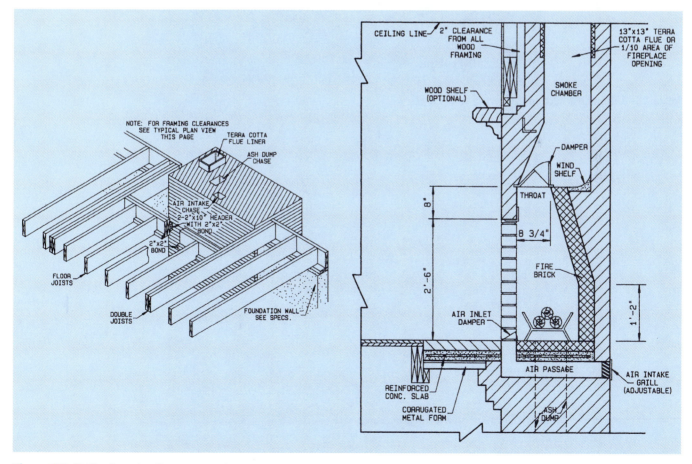

Figure 23–7 Fireplace details. *Courtesy of W. D. Farmer.*

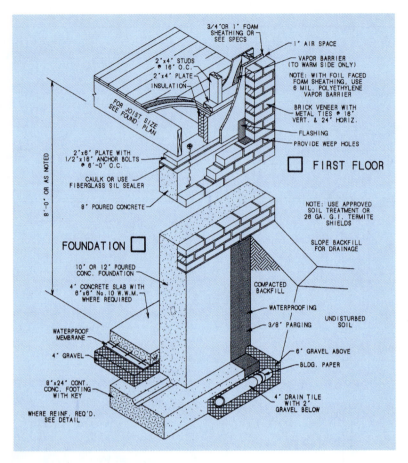

Figure 23–8 **Isometric section.** *Courtesy of W. D. Farmer.*

SB–150SX

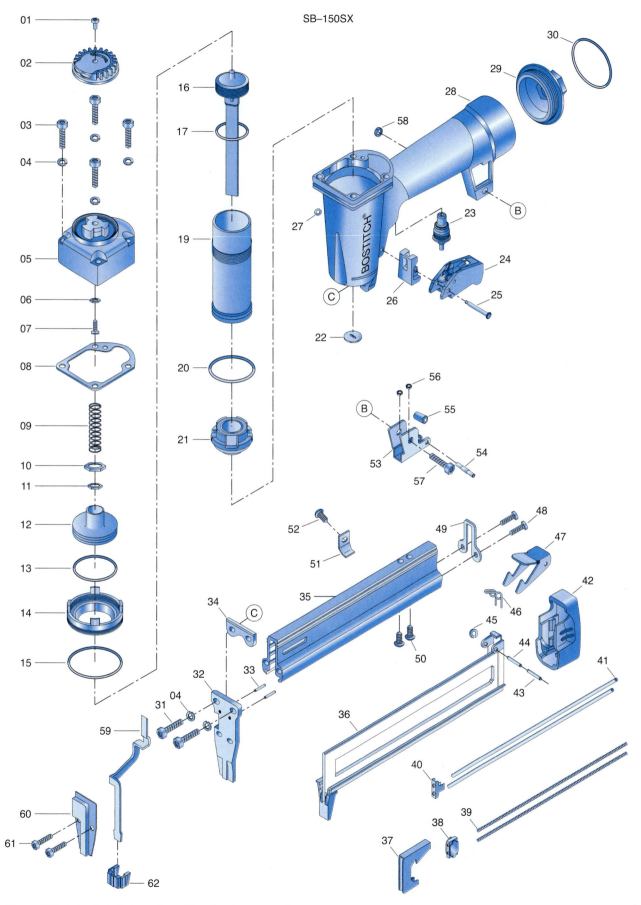

Figure 23–9 Exploded view of a nailer showing internal parts.

Review Questions

1. What is a transverse section view of a house?

2. What is a longitudinal section view of a house?

3. List four important bits of information that are usually given on a typical wall section.

4. What type of drawing is normally drawn at a larger scale than the rest of the drawing set?

5. When looking at a floor plan, how can you tell if a section view exists that shows more detail of an area?

Activities

Refer to the drawings of the house in Appendix B at the end of the book to complete the assignment.

1. What is used to show the detail of a complex design, installation, or product?
2. What kind of section drawing is the Section Thru Kitchen Roof?
3. What kind and size material is to be used for the foundation walls?
4. What is used between the foundation wall and the 2×10 floor joists?
5. What kind and size of insulation is used around the foundation? Is this used on the inside or outside of the foundation?
6. What kind and size of material is to be used on the inside of the frame walls?
7. What is the distance from the kitchen countertop over the dishwasher to the bottom of the wall cabinets?
8. Which wall is the oven on—north, south, east, or west?
9. What size studs are used in a typical wall?
10. What is done in the coat closet to create headroom over the stairs.

Appendix A

Scoring for Occupational Work Ethic Inventory

The following scoring is based on an analysis of over 1,500 successful workers from more than 100 occupations from several different occupational categories. The analysis yielded four subcategories that will help you determine your relative occupational work ethic when compared with an industry average.

The OWEI consists of 50 items that should be answered as truthfully as possible. After completing the inventory, calculate your subcategory scores as shown below. There is no subcategory for item 50; therefore it is not used in your scoring.

1 Add the scores (the small numbers next to each selection) from items 42, 47, 31, 37, 43, 28, 32, 29, 41, 48, 46, 22, 17, 33, 19, and 2. Divide the total by 16 (the number of items in subcategory *Interpersonal Skills*).

Your score for Interpersonal Skills is _____.

2 Add the scores from items 11, 10, 6, 49, 45, 7, 14, 36, 38, 40, 18, 20, 5, 15, 27, and 35. Divide the total by 16 (the number of items in subcategory Initiative).

Your score for Initiative is _____.

3 Add the scores from items 4, 3, 8, 1, 16, 12, and 23. Divide the total by 7 (the number of items in subcategory Being Dependable).

Your score for Being Dependable is _____.

4 Add the scores from items 39, 34, 24, 25, 13, 44, 26, 21, 9, and 30. Divide the total by 10 (the number of items in subcategory Unconditional Acceptance of Duty).

Your score for Unconditional Acceptance of Duty is

_____.

5 List your scores below and compare with industry averages. Determine if you feel you need to take corrective action.

OWEI Subcategory	Your Score	Industry Average	You Should Score	Action Needed to Correct (Yes/No)
Interpersonal Skills		5.80	(Minimum) 4.8	
Initiative		5.66	(Minimum) 4.50	
Being Dependable		6.21	(Minimum) 5.00	
Unconditional Acceptance of Duty		2.51 (Your score should be low)	(Maximum) 4.00	

Appendix B

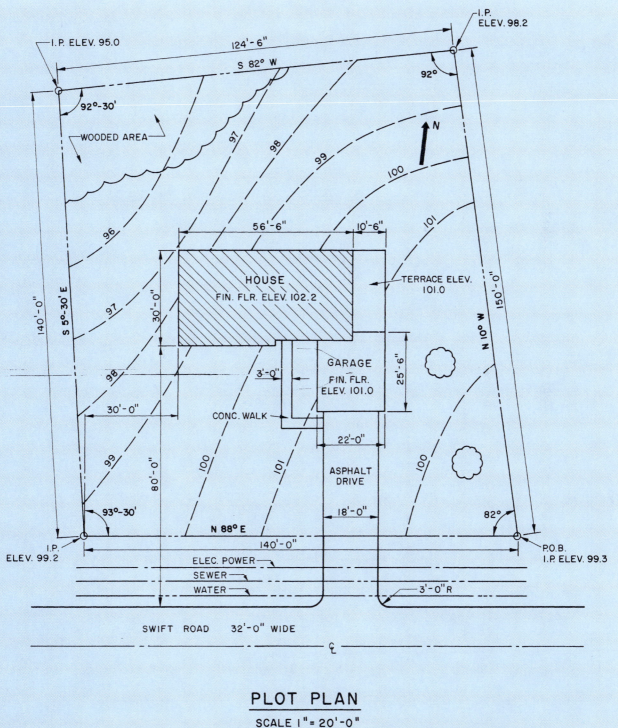

PLOT PLAN

SCALE 1" = 20'-0"

(NOT TO SCALE)

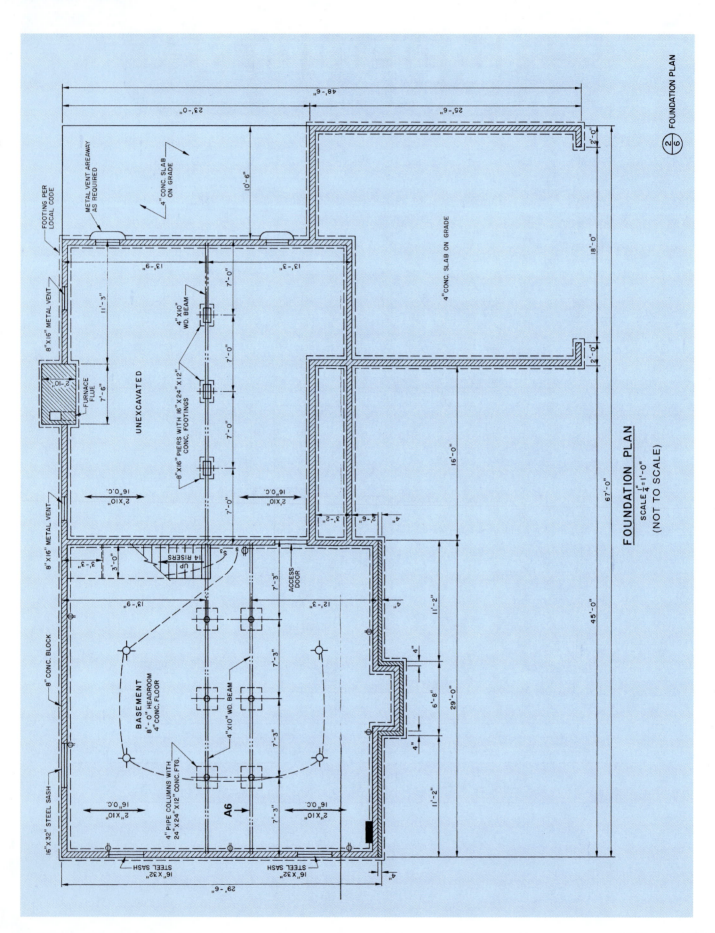

FOUNDATION PLAN
SCALE ¼"=1'-0"
(NOT TO SCALE)

FOUNDATION PLAN
2
6

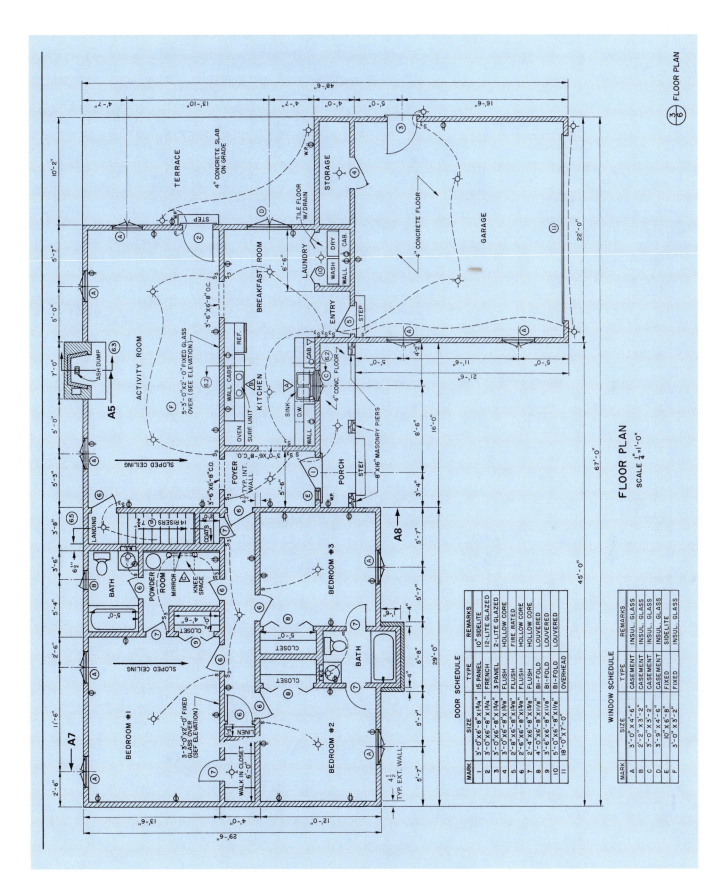

FLOOR PLAN
SCALE ¼"=1'-0"

DOOR SCHEDULE

MARK	SIZE	TYPE	REMARKS
1	3'-0"x6'-8"x1¾"	15 PANEL	10" SIDELITE
2	3'-0"x6'-8"x1¾"	FRENCH	12-LITE GLAZED
3	3'-0"x6'-8"x1¾"	3 PANEL	2-LITE GLAZED
4	3'-0"x6'-8"x1⅜"	FLUSH	HOLLOW CORE
5	3'-0"x6'-8"x1⅜"	FLUSH	FIRE RATED
6	2'-8"x6'-8"x1⅜"	FLUSH	HOLLOW CORE
7	2'-6"x6'-8"x1⅜"	FLUSH	HOLLOW CORE
8	4'-0"x6'-8"x1⅛"	BI-FOLD	LOUVERED
9	3'-6"x6'-8"x1⅛"	BI-FOLD	LOUVERED
10	5'-0"x6'-8"x1⅛"	BI-FOLD	LOUVERED
11	18'-0"x7'-0"	OVERHEAD	

WINDOW SCHEDULE

MARK	SIZE	TYPE	REMARKS
A	3'-0"x4'-6"	CASEMENT	INSUL. GLASS
B	2'-2"x3'-2"	CASEMENT	INSUL. GLASS
C	3'-0"x3'-2"	CASEMENT	INSUL. GLASS
D	3'-9"x4'-6"	CASEMENT	INSUL. GLASS
E	10"x6'-8"	FIXED	SIDELITE
F	3'-0"x3'-2"	FIXED	INSUL. GLASS

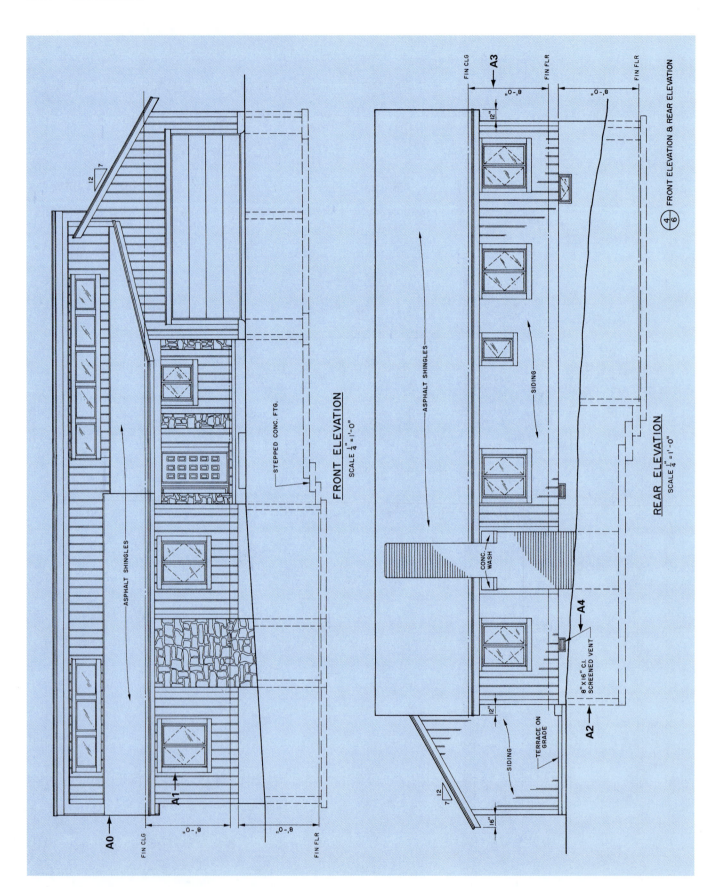

FRONT ELEVATION
SCALE 1/4" = 1'-0"

ASPHALT SHINGLES

STEPPED CONC. FTG.

A0

FIN CLG

8'-0"

FIN FLR

8'-0"

A1

REAR ELEVATION
SCALE 1/4" = 1'-0"

ASPHALT SHINGLES

SIDING

CONC. WASH

SIDING

TERRACE ON GRADE

8" x 16" C.I. SCREENED VENT

A4

A2

12"

16"

12
7

FIN CLG

A3

6'-0"

FIN FLR

8'-0"

FIN FLR

12"

4/6 FRONT ELEVATION & REAR ELEVATION

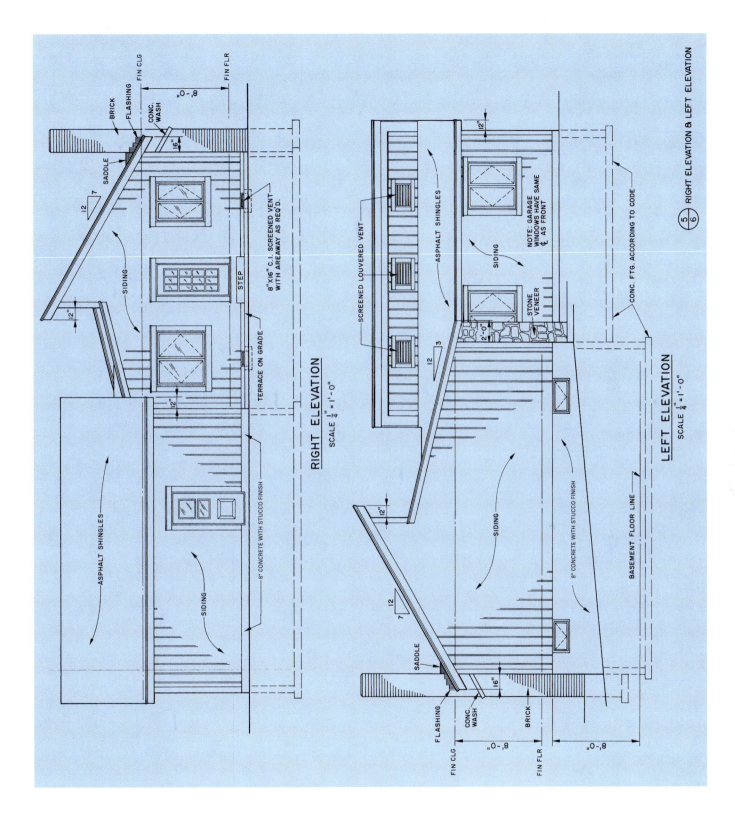

RIGHT ELEVATION
SCALE 1/4" = 1'-0"

BRICK
FLASHING
CONC. WASH
SADDLE
16"
12
7
SIDING
12"
FIN CLG
FIN FLR
8'-0"
8"X16" C.I. SCREENED VENT WITH AREAWAY AS REQ'D.
STEP
TERRACE ON GRADE
12"
ASPHALT SHINGLES
SIDING
8" CONCRETE WITH STUCCO FINISH

LEFT ELEVATION
SCALE 1/4" = 1'-0"

12"
SCREENED LOUVERED VENT
ASPHALT SHINGLES
SIDING
STONE VENEER
NOTE: GARAGE WINDOWS HAVE SAME ₵ AS FRONT
CONC. FTG. ACCORDING TO CODE
12
3
2'-0"
12"
12
7
SADDLE
FLASHING
CONC. WASH
16"
BRICK
FIN CLG
FIN FLR
8'-0"
SIDING
8" CONCRETE WITH STUCCO FINISH
BASEMENT FLOOR LINE
8'-0"

5/6 RIGHT ELEVATION & LEFT ELEVATION

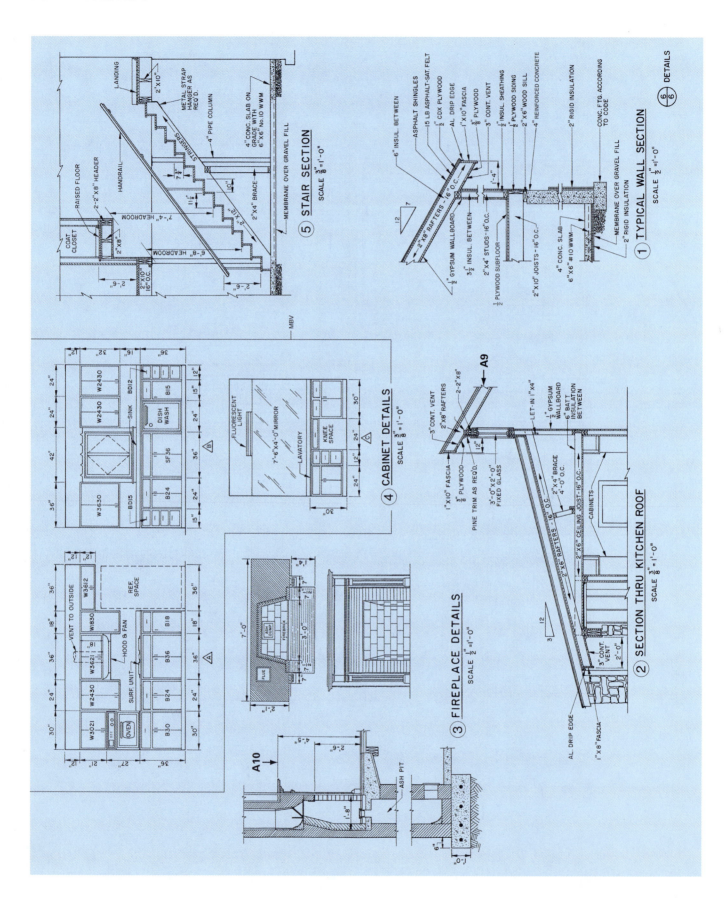

⑤ STAIR SECTION
SCALE 3/8" = 1'-0"

① TYPICAL WALL SECTION
SCALE 1/2" = 1'-0"

⑥ DETAILS
⑥

④ CABINET DETAILS
SCALE 3/8" = 1'-0"

② SECTION THRU KITCHEN ROOF
SCALE 3/8" = 1'-0"

③ FIREPLACE DETAILS
SCALE 1/2" = 1'-0"

Glossary

amount principal plus interest.

cantidad principal más interés.

ampere the unit of measure for electric current. Also abbreviated amp. Many power tool motors are sized according to the amperage their motor draws.

amperio la unidad de la medida para la corriente eléctrica. También abreviado amp. Muchos motores de la herramienta eléctrica se clasifican según el amperaje que su motor dibuja.

anchor a device that can be driven or set in concrete, masonry, or other material to provide a place to attach a bolt. There are several types of anchors.

ancla un dispositivo que se pueda conducir o fijar en el concreto, la albañilería, o el otro material para proporcionar un lugar a la fijación un perno. Hay varios tipos de anclas.

apprentice a person who is being trained to work in the building trades. Apprentices attend classes and work under the supervision of a skilled craftsman.

aprendiz una persona que se está entrenando para trabajar en los comercios de edificio. Los aprendices atienden a clases y al trabajo bajo supervisión de un artesano experto.

area the space inside a shape.

area el espacio dentro de una forma.

ball bearing a style of bearing in which moving parts roll on steel balls.

rodamiento de bolitas un estilo de llevar en que piezas móviles rueden en las bolas de acero.

base of a triangle the side opposite the corner from which the height is measured. This can be any side of the triangle.

base de un triángulo el lateral enfrente de la esquina de la cual se mide la altura. Éste puede ser cualquier lado del triángulo.

blade (rafter square) the longer arm of a square.

lámina (cuadrado del viga) el brazo más largo de un cuadrado.

body language unwritten, unspoken message communicated by the way we hold our body or the expressions on our face.

la lengua de cuerpo el mensaje no escrita y no hablado que nos comunica de la manera llevamos nuestro cuerpo y de la expresión de la cara.

box nail has a thin shank like a finishing nail, but a flat head like a common nail. Usually coated to prevent loosening.

el clavo de la caja tiene una caña fina como un clavo que acaba, pero una cabeza plana como un clavo común. Generalmente cubierto para evitar aflojamiento.

brad a very short nail with a small head, used to fasten thin parts.

alfilercillo un clavo muy corto con una cabeza pequeña, usada para sujetar piezas finas.

cap screw a small bolt, usually with a hexagonal head.

tornillo de casquillo un perno pequeño, generalmente con una cabeza hexagonal.

carriage bolt a large bolt for use in wood. Has a smooth oval head and a section of square shank right below the head.

perno de carro un perno grande para el uso en madera. Tiene una cabeza oval lisa y una sección de la derecha cuadrada de la caña debajo de la cabeza.

centerline a line used to indicate the center axis of an object.

línea central una línea para indicar el eje de centro de un objeto.

chuck the part of a drill that holds the drill bit.

arroje la pieza de un taladro que celebre el pedacito de taladro.

chuck key a special tool used to tighten a drill chuck.

llave de Arroje una herramienta especial usada para apretar una tirada del taladro.

circle a shape in which every point on the perimeter is the same distance from a center point.

circulo una forma en la cual cada punto en el perímetro es a la misma distancia de un punto de centro.

class A fire a fire that involves ordinary materials like paper, cardboard, and wood. Class A fires can be extinguished with water.

fuego de la clase A un fuego que implica los materiales ordinarios como el papel, la cartulina, y la madera. Los fuegos de la clase A se pueden extinguir con agua.

class B fire a fire that involves flammable liquids. Class B fires are extinguished with either dry chemicals or CO_2 (carbon dioxide).

el fuego de la clase B un fuego que implique líquidos inflamables. Los fuegos de la clase B se extinguen con los productos químicos secos o el CO_2 (bióxido de carbono).

class C fire an electrical fire. Class C fires are extinguished with CO_2 (carbon dioxide).

fuego de la clase C un fuego eléctrico. Los fuegos de la clase C se extinguen con el CO_2 (bióxido de carbono).

clinching bending the protruding part of a nail over to make a permanent fastening.

el afianzar doblar la parte que resalta de un clavo encima para hacer una cerradura permanente.

combination blade a saw blade that can be used for ripping and crosscutting.

lámina de la combinación una lámina de sierra que se puede utilizar para rasgar y cortar.

common nail the most common type of nail. Has a heavy, smooth shank and a flat head.

clavo común el tipo más común de clavo. Tiene una caña pesada, lisa y una cabeza plana.

competent person one who is capable of identifying existing and predictable workplace hazards that are unsanitary, hazardous, or dangerous to employees and who has the authorization to take corrective measures to eliminate them.

persona competente el que es capaz de identificar los peligros existentes y fiables del lugar de trabajo que son antihigiénicos, peligrosos, o peligrosos a los empleados y que tiene la autorización de tomar medidas correctivas de eliminarlas.

conductor a material that allows electricity to flow.

conductor un material que permite que fluya la electricidad.

contractor the person who owns the construction business. Contractors enter into contracts with customers to do specified construction work. Contractors hire workers or other subcontractors to complete the contracted work.

contratista la persona que posee el negocio de la construcción. Los contratistas entran en contratos con los clientes para hacer el trabajo de

construcción especificado. Los contratistas emplean a trabajadores o a otros subcontratistas para terminar el trabajo contraído.

corporation a form of business ownership in which people who are not involved in operating the business own shares of the company. The company is operated by a board of directors.

corporación una forma de propiedad del negocio en la cual la gente, que no está implicada en el funcionamiento del negocio, posee las partes de la compañía. Una directiva junta funciona maneja la compañía.

craft see **skilled trades.**

el arte considera comercios expertos.

craft union members of a particular craft who are organized to work for the betterment of all members of the group. Union members pay dues as a requirement of their membership.

unión de arte miembros de un arte particular que se organiza para trabajar para la mejora de todos los miembros del grupo. Los miembros de unión pagan deudas como requisito de su calidad de miembro.

cube a three-dimensional shape in which height, width, and depth are all equal.

cubique una forma tridimensional en la cual la altura, la anchura, y la profundidad estén toda igual.

cutting-plane line a line used to show where a section view is taken from and in what direction it is viewed.

línea del plano secante una línea que demonstra de donde una opinión de la sección se toma y en qué dirección se ve.

decimal fraction a number representing a quantity of less than 1 and expressed according to the decimal number system.

fracción decimal un número que representa una cantidad de menos de 1 y expresada según el sistema de numeración decimal.

decimal point the dot used to indicate the separation between whole numbers and decimal fractions.

el punto decimal que indica la separación entre los números enteros y las fracciones decimales.

deck screw similar to a drywall screw except it is more corrosion-resistant.

tornillo de la cubierta igual que un tornillo del muro seco a menos que sea más resistente a la corrosión.

denominator the number on the bottom of a common fraction.

denominador el número en el fondo de una fracción común.

developer the person or company that buys undeveloped land and works with architects and contractors to develop it into more valuable property.

revelador la persona o la compañía que compran tierra subdesarrollada y trabajan con los arquitectos y los contratistas para desarrollarla en una característica más valiosa.

difference the result of subtraction.

diferencia el resultado de la substracción.

dimension line a line used to indicate the size of an object or feature.

la línea de Dimensione una línea usada para indicar el tamaño de un objeto o de una característica.

double-insulated a style of electric tool construction that shields the user from the electric parts of the tool.

doble-aíslo un estilo de la construcción eléctrica de la herramienta que blinda a usuario de las piezas eléctricas de la herramienta.

drywall screw a light-gauge screw with a Phillips or square-driven head, used for fastening drywall to framing. Drywall screws are not as strong or as corrosion-resistant as other screws.

tornillo de muro seco un tornillo calibre lidero con una cabeza de Phillips o dirigida por cuadrado usado para abrochar al enmarcado. Los tornillos de muro secos no son tan fuertes o tan resistentes a la corrosión como otros tornillos.

duplex nails a common nail with two heads, so that one can be driven tight and the other is still exposed for removal.

el duplas clava un clavo común con dos cabezas, para poder conducir uno firmemente y el otro todavía se exponga para el retiro.

elevation a view of an object as seen from the side and showing the height of the object.

elevación una vista de un objeto según lo visto del lado y de demostrar la altura del objeto.

ethics the discipline dealing with what is good and bad and with moral duty and obligation.

el ética la disciplina que trata de cuál es bueno y malo y con deber y la obligación morales.

extension line a line used to indicate the extent of a dimension line.

línea de extensión una línea usada para indicar el grado de una línea de la dimensión.

4:1 rule for ladders the rule that stipulates that ladders should be 1 foot away from the vertical surface against which they are placed for every 4 feet in height.

regla de 4:1 para las escalas la regla que estipula que las escalas deben ser 1 pie lejos de la superficie vertical contra la cual se colocan para cada 4 pies en altura.

factor a number that when multiplied by another factor produces another number.

el factor un número que cuando es multiplicado por otro factor produce otro número.

finishing nail has a thin shank and small head that can be driven beneath the surface of the wood.

el clavo que acaba tiene una caña fina y una cabeza pequeña que se

puedan conducir debajo de la superficie de la madera.

fire triangle consists of heat, fuel, and oxygen (the three sides of the triangle); the three elements must be present for a fire to burn.

el triángulo del fuego consiste el calor, el combustible, y el oxígeno (los tres lados del triángulo); los tres elementos deben estar presentes para quemar un fuego.

floor plan shows the size and locations of rooms and most features in the rooms or on the specified level of the building.

el plan de piso demuestra el tamaño y las localizaciones de cuartos y de la mayoría de las características en los cuartos o en el nivel especificado del edificio.

foundation plan shows the layout and size of the foundation.

el plan de la fundación demuestra la disposición y el tamaño de la fundación.

fraction bar the horizontal line in a common fraction.

barra de fracción la línea horizontal en una fracción común.

grade line a line on a drawing that shows the surface level of the ground.

la línea de Califique una línea en un dibujo que demuestre el nivel superficial de la tierra.

ground (electrical) a conducting body that serves as the common return path for an electric circuit. A ground typically has zero potential. The earth may also be used as a ground.

molió (eléctrico) un cuerpo que conduce como la trayectoria de vuelta común para un circuito eléctrico. Una tierra tiene típicamente potencial cero. La tierra se puede también utilizar como tierra.

ground fault circuit interrupter (GFCI) a protective device that opens the electric circuit when an imbalance in the amount of current flow between the conductors is sensed.

interruptor de tierra del circuito de la avería (GFCI) un dispositivo protector que abre el circuito eléctrico cuando se detecta un desequilibrio en la cantidad de flujo actual entre los conductores.

hash mark a short line on a drawing, used to indicate the end of a dimension line.

marca de índice una línea corta en un dibujo, usado para indicar el extremo de una línea de la dimensión.

headers the framing across the top of a window or door opening.

jefes el enmarcar a través de la tapa de una ventana o de una abertura de la puerta.

height of a triangle the length of a line drawn perpendicular to one side of a triangle and extending to the opposite corner.

altura de un triángulo la longitud de una línea perpendicular dibujado a un lado de un triángulo y de extender a la esquina opuesta.

hidden line a dashed line used to show edges that are hidden from normal view.

la línea ocultada una línea discontinuada que demostra los bordes que se ocultan de la visión normal.

horseplay practical jokes and playful activity that are inappropriate on a construction site.

payasadas bromas prácticas y actividad juguetona que son inadecuadas en un emplazamiento de la obra.

hypotenuse the side of a right triangle that is opposite the right angle.

hipotenusa el lado de un triángulo derecho que está enfrente del angulo recto.

interest payment for the use of money; expressed as a percentage of the principal. Interest is usually stated.

interés pago de para el uso del dinero; expresado como porcentaje del principal. El interés se indica generalmente.

isometric drawing a drawing with horizontal lines drawn at an angle of 30°.

dibujo isométrico un dibujo con las lineas horizontales dibujadas en ángulo de 30°.

journeyman a skilled craft worker who has completed an apprenticeship or otherwise proved his or her ability in the trade.

oficial trabajador experto del arte que ha terminado un aprendizaje o una su o su capacidad de otra manera probada en el comercio.

kerf the cut made by a saw.

corte el corte hecho por una sierra.

labels user instructions found on most construction products. The product label contains valuable information.

etiquetas las instrucciones del usuario que se encuentra en la mayoría de los productos de la construcción. La etiqueta del producto contiene la información valiosa.

laborer an unskilled or semiskilled worker on a construction site.

trabajador una persona no experta y meno calificada en un emplazamiento de la obra.

lag screw a large wood screw with either a square or hexagonal head.

tornillo del retraso un tornillo de madera grande con una cabeza cuadrada o hexagonal.

lay the direction in which the strands are wound around the core of a wire rope. Also the distance along a rope in which a strand makes a complete turn around the core.

colocación la dirección en la cual los filamentos se hieren alrededor de la base de una cuerda de alambre. También la distancia a lo largo de una cuerda en la cual un filamento hace una vuelta completa alrededor de la base.

leader a line used to associate a label with an object. Leaders usually have arrowheads.

líder una línea asociaba una etiqueta a un objeto. Los líderes tienen generalmente puntas de flecha.

level parallel to the earth's surface.

nivel paralelo a la superficie de la tierra.

longitudinal section a section view showing what would be seen at an imaginary cut through the length of the building or part.

sección longitudinal una demostración de la opinión de la sección qué sería vista en un corte imaginario a través la longitud del edificio o de la parte.

lowest common denominator the smallest denominator that can be used to express all of the common fractions in a set.

el denominador común más bajo el denominador más pequeño que se puede utilizar para expresar todas las fracciones comunes en un sistema.

Material Safety Data Sheet (MSDS) gives complete information about the product and what to do in the event of exposure. An MSDS is required to be available for any substance that might be harmful.

la hoja de datos material de seguridad (MSDS) da la información completa sobre el producto y qué hacer en el acontecimiento de la exposición. Un MSDS se requiere estar disponible para cualquier sustancia que pudiera ser dañosa.

metric system a system of measurement based on 10 (often called the SI system).

sistema métrico un sistema de la medida basado en 10 (a menudo llamado el sistema del SI).

minuend the number from which another number is to be subtracted.

minuendo el número de el cual otro número debe ser restado.

mixed number a number made up of a whole number plus a common fraction.

número mezclado un número compuso de un número entero y una fracción común.

model code a suggested building code, intended to be adopted as is or with revisions to become an official code of a particular government.

código modelo un código de edificio sugerido, previsto para ser adoptado al igual que o con las revisiones a convertirse un código oficial de un gobierno particular.

nominal dimensions the dimensions of a product before allowances or adjustments are made. The sizes of many construction materials are identified by their nominal dimensions. For example, nominal dimensions of lumber are the dimensions of lumber before it is dried and planed or the dimensions of masonry units including the mortar joints.

la dimensiona nominal las dimensiones de un producto antes de que se hagan los permisos o los ajustes. Los tamaños de muchos materiales de construcción son identificados por sus dimensiones nominales. Por ejemplo, las dimensiones nominales de la madera de construcción son las dimensiones de la madera de construcción antes de ella se secan y se acepillan o las dimensiones de las unidades de la albañilería incluyendo los empalmes del mortero.

numerator the number on the top of a common fraction.

numerador el número en la tapa de una fracción común.

object line a heavy solid line used to show the outline and/or shape of an object.

línea de objeto una línea llena pesada usada para demostrar el contorno y/o la forma de un objeto.

oblique drawing a drawing with one surface shown on the plane of the paper and the adjoining surface at an angle.

dibujo oblicuo un dibujo con una superficie demostrada en el plano del papel y la superficie colindante en ángulo.

orienting showing the relationship of one drawing to another.

orientación demostrando la relación de una que dibuja a otra.

orthographic projection a style of drawing in which separate sides of an object are shown as if projected against the inside of a glass box.

proyección ortográfica un estilo del dibujo en que lados separados de un objeto se demuestran como si estén proyectados contra el interior de una caja de cristal.

OSHA refers both to the state and federal Occupational Safety and Health Administration and also to state administrations. OSHA also stands for the Occupational Safety and Health Act, which is administered by the Occupational Safety and Health Administration. OSHA generally refers to the laws that are intended to keep workers safe.

OSHA refiere ambos al estado y federal Administración de Seguridad y Salud Ocupacional y también a las administraciones estado. OSHA también refiere al Acto de la Seguridad y Salud Ocupacional, cual es administrado por la Administración de Seguridad y Salud Ocupational. El OSHA refiere generalmente a los leyes que se piensan para mantener a trabajadores seguros.

partnership a form of business in which more than one person shares the ownership and operating duties for a company.

sociedad una forma de negocio en la cual más de una persona comparte la propiedad y los deberes de funcionamiento para una compañía.

penny size (abbreviated *d*) refers to the size of a nail. Nails are measured by an old system that used the number of pennies to purchase 100 nails of that size. The higher the penny size, the longer the nail.

tamaño del penique (d abreviada) refiere al tamaño de un clavo. Los clavos son medidos por un viejo sistema que utilizó el número de peniques para comprar 100 clavos de ese tamaño. Cuanto más alto es el tamaño del penique, más largo es el clavo.

percentage (percent—%) hundredths of the whole.

porcentaje (por ciento—%) centésimo del conjunto.

perimeter the distance around the outside of a shape.

perímetro la distancia alrededor del exterior de una forma.

personal protective equipment (PPE) any safety equipment you wear to protect yourself from safety hazards.

equipo protector personal (PPE) cualquier equipo de seguridad que usted use para protegerse contra peligros de seguridad.

pitch a measurement of the number of threads in 1 inch of a screw or bolt.

eche una medida del número de hilos de rosca en 1 pulgada de un tornillo o de un perno.

pitch (saw) the coarseness of the teeth of a saw. Pitch is measured in points per inch.

eche (sierra) la tosquedad de los dientes de una sierra. La echada se mide en puntos por pulgada.

plan view a view of an object as seen from directly above.

vista de plan una vista de un objeto según lo visto de directamente arriba.

plate the uppermost framing member in the wall.

platee el miembro que enmarca más supremo de la pared.

plumb perfectly perpendicular to the earth's surface.

sondee perfectamente perpendicular a la superficie de la tierra.

plunge cut a cut made by plunging the saw in the middle of the work piece instead of cutting in from an outside edge.

hunda el corte un corte hecho hundiendo la sierra en el centro del pedazo del trabajo en vez del corte adentro de un borde exterior.

pneumatic tools tools powered by compressed air.

herramientas neumáticas herramientas accionadas por el aire comprimido.

polarized plug an electric plug having one prong wider than the other so that it can only be plugged into the receptacle one way.

enchufe polarizado un enchufe eléctrico que tiene un diente más de par en par que el otro para poderlo tapar solamente en una forma del receptáculo.

portable ladder a ladder that can easily be picked up and carried to another location.

escala portable una escala que se puede tomar fácilmente y llevar a otra localización.

principal the amount of money invested or borrowed.

principal la cantidad de dinero invertida o prestada.

profession an occupation that requires more than four years of college and a license to practice.

profesión una ocupación que requiere más de cuatro años de la universidad y de una licencia de practicar.

pump jack a device that attaches to a vertical pole and can be pumped up and down the pole. Pump jacks are used to support planks on which workers stand.

bombee el gato un dispositivo que une a un poste vertical y se pueda bombear arriba y abajo del poste. Los gatos de la bomba se utilizan para apoyar los tablones en los cuales los trabajadores están parados.

pythagorean theorem a mathematical law that says the sum of the squares of the sides of a right triangle are equal to the square of the hypotenuse.

el teorema piíta gorrean una ley matemática que diga la suma de los cuadrados de los lados de un triángulo derecho es igual al cuadrado de la hipotenusa.

rate percentage at which interest is calculated.

la tarifa porcentaje del cual se calcula el interés.

rated capacity the amount of weight the manufacturer has specified for the maximum load on a sling.

capacidad clasificada la cantidad de peso que el fabricante ha especificado para la carga máxima en una honda.

reciprocate move back and forth in a straight line. Some saw blades reciprocate.

intercambiar el movimiento hacia adelante y hacia atrás en una línea recta. Algunas láminas de sierra intercambian.

rectangle a shape with four 90° corners and with opposite sides of equal length.

rectángulo una forma con cuatro esquinas del 90° y con los lados opuestos de la longitud igual.

reference marks callouts on plans and elevations that indicate where details or sections of important features have been drawn.

marcas de referencia planes y las elevaciones que indican donde han estado exhaustos los detalles o las secciones de características importantes.

reversing switch a switch found on most electric drills that allows the user to reverse the direction of the drill.

invertir el interruptor un interruptor encontrado en los taladros más eléctricos que permite que el usuario invierta la dirección del taladro.

right angle a 90° angle.

ángulo recto un ángulo del 90°.

Right to Know rule the OSHA rule that says that every worker has a right to know about any substances on the job that might be harmful to humans.

derecho de saber regla la regla del OSHA que dice que cada trabajador tiene una derecha de saber sobre cualquier sustancia en el trabajo que pudo ser dañoso a los seres humanos.

right triangle a triangle with a 90° angle.

triángulo derecho un triángulo con un ángulo del 90°.

roller bearing a style of bearing in which moving parts roll on small steel rollers.

cojinete de rodillo un estilo de llevar en qué piezas móviles ruedan en los rodillos de acero pequeños.

scaffolds temporary work platforms.

plataformas trabajo temporal de los andamios.

scale the amount of reduction or enlargement of an object that is drawn other than the actual size. Also a device used to measure objects that are drawn proportionately smaller or larger.

escala la cantidad de reducción o de ampliación de un objeto que se dibuje con excepción del tamaño real. También a el dispositivo medía los objetos que son proporcionalmente más pequeños o más grandes dibujado.

screw gauge a number representing the thickness of a screw. The higher the gauge, the thicker the screw.

galga de tornillo un número que representa el grueso de un tornillo. Cuanto más alta es la galga, más grueso es el tornillo.

section a view of an object as though the object had been cut in half.

sección una vista de un objeto como si el objeto había sido cortado adentro a medias.

semiskilled labor workers with very limited training or skills in the construction trades.

trabajadores menos calificados con el entrenamiento o habilidades muy limitado en los comercios de la construcción.

sheet metal screw usually a self-tapping screw, used for fastening sheet metal.

tornillo del metal de hoja generalmente un tornillo, usado para el metal de hoja de la cerradura.

shoe as used on power tools, the part of the tool that rests on the work piece.

zapato según lo utilizado en las herramientas eléctricas, la pieza de la herramienta esa restos en el pedazo del trabajo.

sides of a right triangle the two sides next to the right angle. The hypotenuse is not referred to as a side.

lados de un triángulo derecho los dos lados al lado del ángulo recto. La hipotenusa no se refiere como lado.

site plan also called plot plan. Shows the size of the building site and the location of the building, driveways, sidewalks, and utilities on the site.

plan de sitio también llamado el diagrama del plan de sitio. Demuestra el tamaño del sitio de edificio y de la localización del edificio, de las calzadas, de las aceras, y de las utilidades en el sitio.

Skil saw a term sometimes used to mean a portable circular saw. The first portable circular saw was made by Skil, and so some people call all portable circular saws Skil saws.

sierra de Skil una palabra usada a veces para significar un sierra circular portable. El primer sierra circular portable fue hecho por Skil, y por eso algunas personas llaman todos los sierras circulares portables Sierras Skil.

skilled trades the building trades—carpenters, electricians, plumbers, painters, and so on. These occupations require training and skill. The skilled trades are often referred to as the crafts.

comercios expertos los carpinteros de los comercios de edificio, electricistas, fontaneros, pintores, etcétera. Estas ocupaciones requieren el entrenamiento y la habilidad. Los comercios expertos se refieren a menudo como los artes.

sleeve bearing a style of bearing in which moving parts ride on a smooth metal sleeve or tube.

cojinete de manga un estilo de llevar en qué piezas móviles montan en una manga o un tubo lisa del metal.

sole proprietorship a business whose owner and operator are the same person.

propiedad única un negocio que dueño y operador es la misma persona.

solid a three-dimensional shape. Spheres and cubes are two types of solids.

sólido una forma tridimensional. Las esferas y los cubos son dos tipos de sólidos.

square a shape with four 90° corners and four equal-length sides. Square also refers to the result of multiplying a number by itself.

ajuste una forma con cuatro esquinas del 90° y cuatro lados de la igual-longitud. El cuadrado también refiere al resultado de multiplicar un número por sí mismo.

stove bolt a small bolt with a round or flat head and fitted for a screwdriver.

perno de la estufa un perno pequeño con una cabeza redonda o plana y cabido para un destornillador.

subcontractor a contractor who is performing work for another contractor.

subcontratista un contratista que está realizando el trabajo para otro contratista.

subtrahend a number that is to be subtracted from another number.

menos un número que debe ser restado de otro número.

sum the result of addition.

suma el resultado de la adición.

tag line a light line attached to a load to control it by hand.

línea de marca una línea ligera unida a una carga para controlarla a mano.

tapcon a concrete screw.

tapcon un tornillo concreto.

target rod a graduated pole used with a builder's level to measure elevation.

apuntar la barra un poste graduado utilizado con el nivel de un constructor medir la elevación.

technicians technicians provide a link between the skilled trades and the professions by using mathematics, computer skills, specialized equipment, and knowledge of construction.

los técnicos técnicos proporcionan un acoplamiento entre los comercios expertos y las profesiones usando matemáticas, habilidades de la computadora, el equipo especializado, y el conocimiento de la construcción.

thimble a steel insert in the eye of a wire rope sling to prevent kinking and wear.

dedal un relleno de acero en el ojo de una honda de la cuerda de alambre a evitar el enroscarse y a usar.

toenailing driving a nail at an angle into the face of one piece to hold it to another piece. Toenailing usually requires at least one nail on each side of the piece being toenailed.

el clavar oblicuamente conducir un clavo en ángulo en la cara de la una sola pieza para sostenerla a otro pedazo. El clavar oblicuamente requiere generalmente por lo menos un clavo en cada lado del pedazo que es clavado oblicuamente.

tongue (rafter square) the shorter arm of a square.

machihembre (cuadrado de la viga) el brazo más corto de un cuadrado.

transverse section a section view showing what would be seen at an imaginary cut across the building or part.

sección transversal una demostración de la opinión de la sección qué sería

vista en un corte imaginario a través del edificio o de la parte.

triangle a shape formed by three sides.

triángulo una forma formada por tres lados.

tripod three-legged stand for holding a builder's level or a transit.

trípode Soporte de tres piernas para tener un nivel del constructor o un tránsito.

typical section a section view that shows typical construction. Variations of this detail would not be typical and would require special detail drawings.

sección típica una opinión de la sección que demuestra la construcción típica. Las variaciones de este detalle no serían típicas y requerirían dibujos de detalle especiales.

U.S. customary system the system of measurements used in the United States based on inches, feet, quarts, gallons, pounds, and so on.

sistema acostumbrado de Estados Unidos el sistema de las medidas usadas en los Estados Unidos basados en las pulgadas, pies, cuartos de galón, galones, libras, etcétera.

unskilled labor workers with no specific training in the construction trades. This term also applies to work that does not require training.

los trabajadores del trabajo inexperto sin el entrenamiento específico en la construcción negocian. Este término también se aplica al trabajo que no requiere el entrenamiento.

values what we believe is right and wrong.

los valores que creemos correctos e incorrectos.

voltage the electrical pressure that causes current to flow, measured in volts (sometimes abbreviated as V). Voltage is sometimes called electromotive force (EMF) because it is the force that causes electrons to move. Voltage also represents the difference of potential, or potential difference, in a circuit.

voltaje la presión eléctrica que hace la corriente fluir, medida en voltios (abreviados a veces como V). El voltaje a veces se llama la fuerza electromotriz (EMF) porque es la fuerza que hace electrones moverse. El voltaje también representa la diferencia del potencial, o la diferencia potencial, en un circuito.

volume the space enclosed by a three-dimensional figure.

volumen el espacio incluido por una figura dimensional tres.

wire rope clip a clamping device used to hold two pieces of wire rope to form an eye.

clip de la cuerda de alambre un dispositivo que afianzaba con abrazadera llevaba a cabo dos pedazos de la cuerda de alambre para formar un ojo.

wire rope a rope made from strands of wire wrapped around a core in a particular way. Wire rope is sometimes incorrectly called cable.

cuerda de alambre una cuerda hecha de los filamentos del alambre envueltos alrededor de una base de una manera particular. La cuerda de alambre a veces incorrectamente se llama cable.

work practices the things a worker does and how he or she works—these practices have a lot to do with safety.

el trabajo practica las cosas que un trabajador hace y cómo él o ella trabajar-estas prácticas tiene mucho hacer con seguridad.

working conditions the things in the work environment that affect your work.

condiciones de trabajo las cosas en el ambiente del trabajo que afectan su trabajo.

Index

Note: Page numbers in **bold** indicate a table or figure.